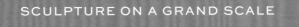

SCULPTURE ON A GRAND SCALE

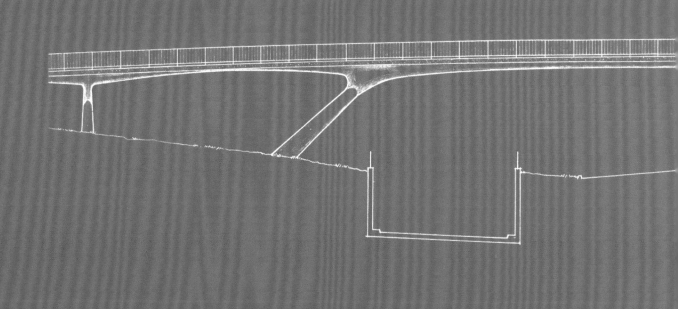

SCULPTURE ON A GRAND SCALE

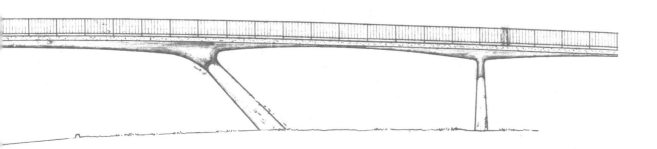

JACK CHRISTIANSEN'S THIN SHELL MODERNISM

TYLER SPRAGUE

A MICHAEL J. REPASS BOOK

UNIVERSITY OF WASHINGTON
Seattle

Sculpture on a Grand Scale was published with the assistance of a grant from the Michael J. Repass Book Fund, which supports publications about the history and culture of Washington, Oregon, and Idaho.

Additional support was provided by the University of Washington Architecture Publications Fund.

Printed and bound in China
Design by Katrina Noble
Composed in Miller Text, typeface designed by Matthew Carter
23 22 21 20 19   5 4 3 2 1

UNIVERSITY OF WASHINGTON PRESS
www.washington.edu/uwpress

LIBRARY OF CONGRESS CATALOGING-IN-PUBLICATION DATA ON FILE
LC record available at https://lccn.loc.gov/2018047477

ISBN 978-0-295-74561-9 (hardcover)
ISBN 978-0-295-74562-6 (ebook)

The paper used in this publication is acid free and meets the minimum requirements of American National Standard for Information Sciences—Permanence of Paper for Printed Library Materials, ANSI z39.48-1984.∞

TO RACHEL

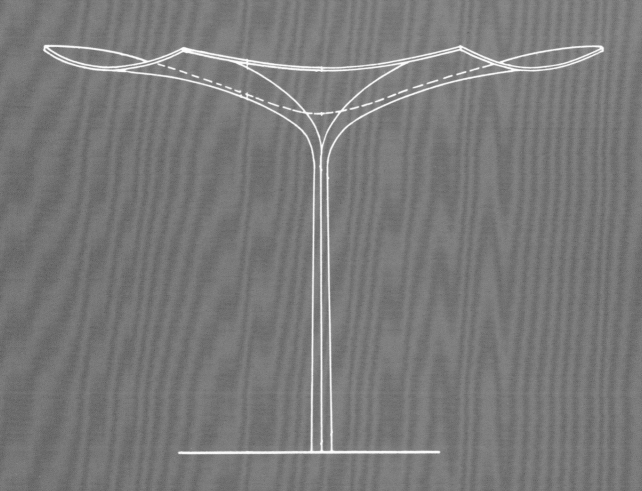

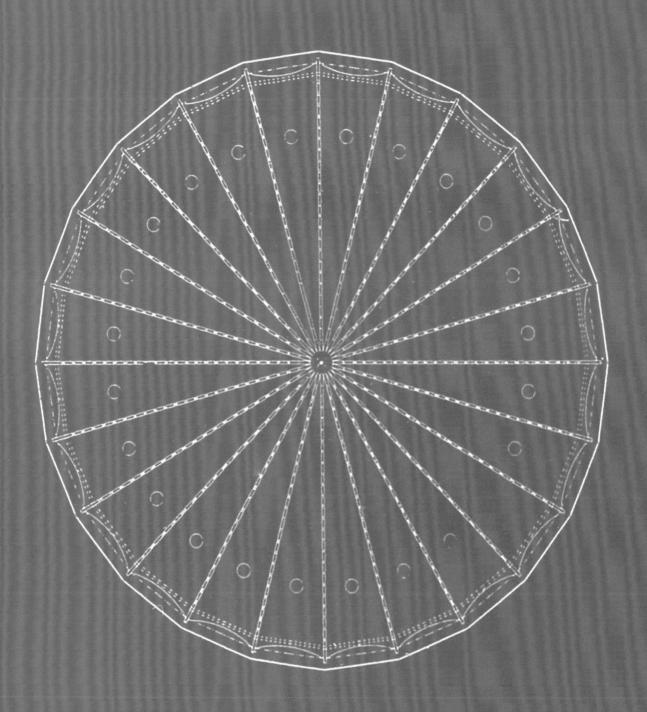

CONTENTS

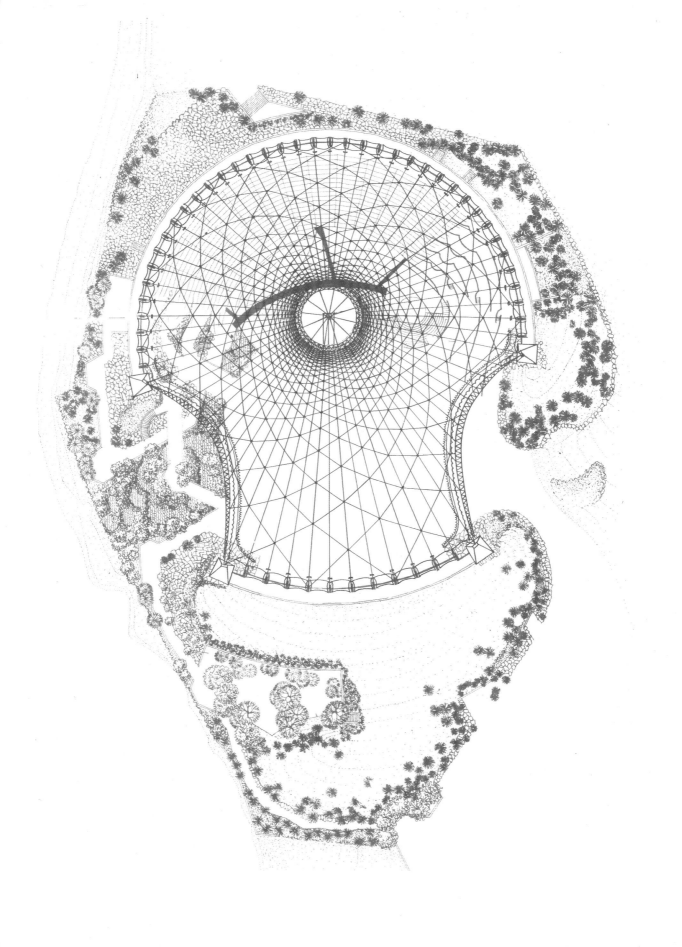

PREFACE

I first learned of Jack Christiansen as a young structural engineer practicing in Seattle. In 2008, I was working in the offices of Magnusson Klemencic Associates, the legacy engineering firm previously named Skilling, Helle, Christiansen & Robertson—where Christiansen had been a partner for over twenty years. Within the office, I heard stories of his thin shell designs and his prominent role as the designer of the by-then demolished Seattle Kingdome—formerly the largest freestanding concrete dome in the world. But the stories were few, and incomplete, raising more questions than they answered. Having arrived in Seattle in 2004, I was unable to visit the Kingdome before its dramatic, nationally televised implosion in 2000—further shrouding Christiansen's work in mystery.

I recalled very little discussion of North American thin shell concrete work in my previous education in modern architectural history. Widespread in Latin America and Europe, thin shell concrete construction, I understood, had been limited in the postwar United States due to building code and labor restrictions, and thus had been less significant. But Christiansen's body of work—anchored by the largest thin shell structure ever built—seemed to contradict this supposition.

Inspired by these and other questions about Christiansen's career, I began my research in 2012. Existing scholarship on Christiansen was limited. A 2006 article in a University of Washington Department of Architecture journal, *Column 5*, by Rainer Metzger, presented several significant projects but only referenced a larger body of work.[1] More significantly, in 2008, Edward M. Segal produced a master's thesis on Christiansen's thin concrete shell designs in the Department of Civil and Environmental Engineering at Princeton University, under the direction of Professor David Billington. This thesis focused primarily on the structural analysis of a few signature works—including the Kingdome—but stopped short of defining the larger historical project.[2] Segal and Billington further recognized Christiansen in a 2011 publication on the fiftieth anniversary of the International Association for Shell and

Spatial Structures (IASS) as an American designer who had pushed the scale and span of thin shell concrete further than Félix Candela and other global designers.[3] That article began to describe the unique role Christiansen played in the larger history of thin shell design. In 2009, architect Susan Boyle interviewed Christiansen as part of an oral history project for the architectural preservation organization Docomomo WEWA.[4] This interview indicated not only Christiansen's prolific career as a structural designer but also his central role in the midcentury modern architecture of the Pacific Northwest.

With support from the University of Washington Center for the Study of the Pacific Northwest in 2012, and drawing from a rough list of projects, I was able to travel to many of Christiansen's extant thin shell concrete project sites throughout Washington State.[5] Seeing his work firsthand, I was even more excited by the physical presence and the scope of Christiansen's work. In 2013, I took part in a documentary film highlighting the contributions of structural engineers to the 1962 Seattle World's Fair, produced by the Structural Engineers Foundation of Washington.[6] This film provided my first opportunity to meet and interview Christiansen in person. Over 6 feet tall, with an imposing frame, he was thoughtful, kind, and deeply humble—in spite of his impressive body of work. From this point on, I did several additional interviews with Christiansen—both at my office at the University of Washington and at Christiansen's thin shell concrete home on Bainbridge Island. Archival research complemented these interviews, including exploration of building plans and slides held by Magnusson Klemencic Associates, study of architectural drawings in Special Collections at the University of Washington Libraries, and a review of Christiansen's personal records.

Through this research, I began to see the Kingdome as a culmination of Christiansen's career trajectory—consisting of over forty years of structural design in the Pacific Northwest. Working as a consulting structural engineer from the early 1950s to the 1990s (a period of significant change), Christiansen practiced alongside the most celebrated architects in the region, with his structural designs enabling some of their most daring and innovative work. It quickly became clear that Christiansen—as a consulting structural engineer—was a key creative contributor to modern architecture in the region.

Although the majority of Christiansen's design work was carried out in tandem with architects, I found that his projects presented a clear trajectory of their own. Independent of the associated architects, Christiansen's sequence of shell designs shows a process of experimentation, expression, and refinement that was indisputably his own. While certainly providing the technical/engineering expertise to execute each work, these designs clearly demonstrate Christiansen's creative influence on the architects with whom he worked—encouraging them to push boundaries of structural and spatial form. These designs often display Christiansen's own personal structural design aesthetic, a position that remained remarkably constant throughout his career, inseparable from his love of international travel and mountaineering. In this way, Christiansen can be credited as a designer in his own right, and with bringing thin shell concrete to the Pacific Northwest.

Jack Christiansen passed away on August 16, 2017, while the manuscript for this publication was being completed. Christiansen was able to review a near-final version before he died, making small, factual corrections but generally approving of the broad story lines and narrative. Humbled about the prospect of a book on his life, Christiansen took this project as an opportunity to reflect and remember the challenges and rewards across his remarkable career. It is my hope that this publication can be a fitting tribute, with his work serving as an inspiration to others.

I am grateful to the many colleagues, family members, and friends who advised and encouraged this work. Professors Jeffrey Karl Ochsner, Vikramaditya Prakash, Meredith Clausen, Bruce Hevly, and Alex Anderson have provided essential guidance in my education and career trajectory. I have enjoyed the support of David Miller and Brian McLaren as successive chairs in the Department of Architecture at the University of Washington. Through the Construction History Society of America, the IASS, and other organizations, I have appreciated the assistance of Thomas Leslie, Marci Uihlein, Robert Whitehead, John Abel, and Juan Ignacio del Cueto. John Ochsendorf, as a teaching collaborator during his time at the University of Washington, greatly enhanced this work through his comments and insight. As Christiansen's work requires significant preservation advocacy, I'm grateful for the efforts of Susan Boyle, Eugenia Woo, Andy Phillips, and the other board members of Docomomo WEWA. I'm thankful to the current leadership of Magnusson Klemencic Associates—including Jay Taylor, Julie Jackson, Ron Klemencic, and Jon Magnusson—who provided access to their firm's archives and understand the importance of the firm's engineering legacy. The librarians and archivists at the University of Washington provided valuable assistance, specifically Alan Michaelson and Noreen Jacky in the College of Built Environments Library and Nicolette Bromberg and Kelly Daviduke in the University of Washington Special Collections Library.

I am indebted to the Christiansen family for their openness and to Jack Christiansen himself, for the wonderful memories he shared.

I am also thankful to my family for their unwavering support—specifically my two daughters, Violet and Eloise, and my loving wife, Rachel.

SCULPTURE ON A GRAND SCALE

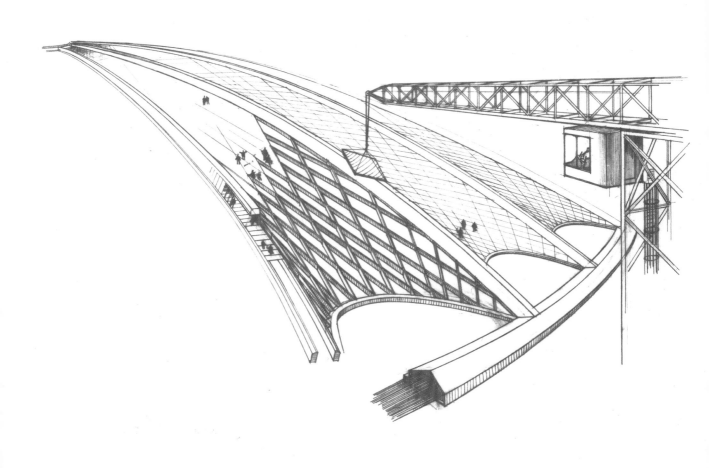

INTRODUCTION

Uncovering an American Shell Designer

To be a structural engineer, I think, is a lot of fun—because of the finished product. The finished product is a great big thing and you can see it! Walk around it! And it makes you like a sculptor, creating sculpture on a grand scale. And of course, to deal with the forces and stresses, well, that's a lot of fun.

—JACK CHRISTIANSEN, 2009

ON MARCH 27, 1976, the Seattle Kingdome opened with great fanfare. Over fifty-four thousand people filled the cavernous volume of the multipurpose stadium to celebrate Seattle's newest civic resource. This housewarming party for the Kingdome had parades of Boy Scouts and Girl Scouts, logrollers, drill teams, marching bands, antique cars, singers, dancers, and Rufus the Frisbee-catching dog. Fireworks exploded overhead. Promoter Tommy Walker claimed, "For this event, we are assembling under one roof the largest cast of performers in the history of Seattle."[1]

The completion of the Kingdome—formally named the King County Multipurpose Stadium—was indeed cause for celebration. After years of political and economic turmoil, the region finally had a singular building that could accommodate its expanding civic ambitions. Municipally owned, the Kingdome was instrumental in bringing new professional soccer, baseball, and football teams to the Pacific Northwest, giving Seattle a place in the national sports conversation. High-profile musicians and performers could now stage large-scale concerts. Art exhibitions, trade shows, and cultural festivals also had a venue to serve Seattle's growing population. The massive, 9-acre roof—a thin shell concrete dome—enabled the stadium to host events at any time of year, in spite of the persistent rainy weather. The Kingdome became a unifying civic form, gathering people and activities from around the region and advancing Seattle's presence on a national stage. The *Seattle Post-Intelligencer* declared 1976 as the "Year of the Dome."[2]

Amid the festivities, John V. (Jack) Christiansen watched from the stadium floor. Few people at the grand opening recognized the tall engineer among them who was uniquely responsible for the design of the dome overhead. In the realization of the Kingdome, Christiansen had exceeded his role as structural engineer, guiding the project and its design through severe legal challenges and budget restrictions. Relying on over twenty years of experience, Christiansen not only had created the largest concrete dome in the world—a remarkable feat of structural engineering—but had done so with a design that could be built at an extremely low cost. His ability to accommodate the effects of drastically inflating construction expenses had made the Kingdome a reality. Standing in the middle of the vaulted space, Christiansen took pride in the multipurpose utility of the Kingdome and the role it would play in Seattle's future.

Christiansen had envisioned the Kingdome lasting over a thousand years, only to see it demolished in less than twenty-five. On March 26, 2000, thousands of small explosive devices were detonated in sequence to bring down the concrete dome in a matter of seconds. The concrete crumpled and fell over 200 feet, resembling a crushed eggshell on the ground below. Employees of businesses in nearby Pioneer Square watched from their rooftops, and local news stations broadcast the demolition live. Within seconds, the massive volume was gone.

After years of mounting maintenance debt, conflicting viability assessments, and the threatened departures of both the baseball and the football teams, King County voters had chosen to demolish the Kingdome. Despite the lack of any structural deficiencies, team owners considered the Kingdome no longer suitable for the changing world of professional sports. The new, baseball-only Safeco Field had just been completed the year before as an open-air facility with a retractable roof. Plans were in place for another open-air football and soccer stadium to take the Kingdome's place. A massive, low-rise exhibition hall was slated to fill the space between them. The Kingdome would soon be replaced by these three separate facilities, with new amenities and capabilities that would bring forth the next era in Seattle sports.

Jack Christiansen had been invited to the demolition of the Kingdome. The demolition company had offered him a "front row" seat to the destructive spectacle, but he forcefully declined. When called for his help in positioning explosive devices to most effectively bring down the concrete dome, Christiansen hung up the phone. After years of petitioning the impending demolition and arguing that the new construction was an inappropriate use of civic funds, Christiansen had reluctantly acknowledged the changing condition of sports and entertainment in Seattle and the need for new facilities. He maintained, however, that the stadium still had usefulness to King County and the people of Puget Sound, and demolition would simply be a colossal waste. As the dome fell, Christiansen was out of town on vacation, finding solace and peace in the outdoors.

The construction and then the demolition of the Kingdome were seminal moments in the history of the Pacific Northwest. These equally dramatic changes to the built environment signaled the larger transitions of economics, politics, sports, and culture that were taking place at particular moments in Seattle. Each event was preceded by years of public discussions, controversies, and changing political agendas, with the opening and the demolition of the Kingdome each punctuating a new reality to come. While many people are familiar with the fate of the Seattle Kingdome, fewer recognize the central role that one man, Jack Christiansen, played in its creation and the personal connection he had to its design. Christiansen's personal story is embedded in the Kingdome saga. Its place in the history of the Pacific Northwest is emblematic of Christiansen's entire career and the prominent role he played in shaping the buildings of the region.

Christiansen had designed the Kingdome in the midst of a prolific, world-class career as a structural engineer. Between the early 1950s and the end of his career in the late 1990s, Christiansen designed over a hundred projects in the Pacific Northwest. He designed for the prosperity of the immediate postwar era, the growth of Boeing, and the excitement of the Space Age, through the recessions of the 1970s and the technology boom of the 1990s. Thousands of students in the Pacific Northwest went to school under thin shell concrete vaults designed by Christiansen. Church congregations held services inside spaces shaped by Christiansen's creative structural forms. At the 1962 World's Fair, Seattle's showcase to the world, Christiansen designed expressive structures for multiple buildings that became symbolic of the city's national image. When King County required the largest concrete dome in the world to bring professional sports to the Northwest, Christiansen created the Kingdome. In each era, Christiansen produced innovative structural forms that were celebrated around the world, with profound local impact.

Jack Christiansen's career was also remarkable for the way he achieved this impact—as a creative structural engineer. Structural engineers are not commonly recognized as primary actors in the production of modern buildings, often overshadowed by architects or contractors in the public view. In their work, they are commonly seen as either simply executing the desires of an institutional or architectural client or applying industry-standard solutions to specific buildings. Whatever innovation an engineer may accomplish often remains a technical solution to a problem rather than a creative, individual act of design. Christiansen's work certainly defies this characterization, with designs that were technical and creative, practical and artistic, and always suited to a particular time and place. Christiansen worked closely with architects in a collaborative manner, but also took a lead creative role in defining structural form and space. In his design of safe, freestanding structures, Christiansen never lost sight of the spatial, artistic quality of his creations or of the practicalities of construction necessary to make them real. His structurally expressive designs, with an identifiable signature all his own, became a distinct feature of the midcentury period.

In this work, Christiansen's designs connect him to a global community of creative structural engineers. In his education, Christiansen was exposed to the work of Robert Maillart, Eduardo Torroja, and Pier Luigi Nervi and admired their ability to balance the technical and artistic qualities of structural design. As he began his design career, Christiansen also closely tracked the work of Anton Tedesko and Félix Candela—understanding their unique approaches. The example of these world-famous engineers gave Christiansen the drive to explore structural design in his own way—at times following their example and other times breaking new ground of his own. Early in his career, Christiansen became an American pioneer in incorporating prestressing technology into his shell structures. With the ambition to extend the scale of his thin shell structures, Christiansen achieved larger spans than anyone else in the world and gained international recognition. The Seattle Kingdome was significant not only in the Pacific Northwest; it also held the title of the largest freestanding thin shell concrete dome in the world. As Christiansen's career matured, he began to take his place alongside the structural engineers he admired so deeply—working together on committees, editing publications, and attending conferences—all while continuing to design his own structures. Christiansen's career also outlasted most of his mentors' and colleagues'. He and Swiss engineer Heinz Isler (1926–2009) were two of the few people who continued to design concrete shells into the 1990s.[3]

THIN SHELL CONCRETE

Essential to Christiansen's impact was his work in a specific structural type—thin shell concrete. With a long history in the modern era, thin shell concrete structures are distinct within the broader range of possible structural systems for buildings and offer unique, expressive possibilities in architectural form.

Thin shell concrete involves both a distinct structural form (thin shell) and a structural material (concrete) brought to their limits. While concrete (a basic mix of cement, water, sand, and aggregate) is quite common in building construction, its use in thin shell structures—structures where the thickness is extremely small when compared to the span (often defined as a ratio of 1:20 or greater)—is less so. When this structural form and structural material are combined together, thin shell concrete offers definite advantages over other systems but also demands other careful considerations.

With such thinness, shell structures gain strength from their precise, overall geometrical form. Shell structures require curvature in one or more directions to remain stable. When these are shaped properly, forces move within the shell as almost pure tension or compression, in a structural action generally referred to as shell, or membrane, behavior. Shells must also carry these forces without inducing bending or buckling of the shell surface. Such clarity in their structural behavior enables shell structures to enclose extremely large volumes of space with only a minimal amount of material. When shaped properly, thin shell concrete structures can be extremely materially efficient in their resistance of structural forces.

In a drive to do more with less, Christiansen saw shell structures as a way to get the maximum use out of a small amount of material. As an avid outdoorsman and mountaineer, Christiansen advocated for resource conservation and a careful management of the consumable materials of the earth. This small amount of material also had the potential for an extreme cost efficiency in shell structures, if managed correctly, lowering the overall expense of creating buildings. He also recognized that concrete, as a building material, was widely available and could often be easily produced locally, reducing manufacturing demands and transportation costs. As a modernist, he preferred the appearance of thin, clean lines, smooth surfaces, and wide-open spaces in architecture. Shell structures in concrete satisfied all of these orientations.

This same structural shell also defines the exterior profile of a building and shapes its interior space. The geometry of a building must be able to resist the physical forces of gravity, but it must also create enough usable space to suit the building's function, ensure proper head height clearances, and give the building its external, visible shape. Thus, the act of determining the precise form of a thin shell concrete structure is a process that must balance engineering and architectural priorities, with a design that falls at the boundary of the two professional worlds.

Christiansen embraced the pronounced, formal presence of his shell structures as an architecture all their own. Trained as an architectural engineer, he believed that structural form should be visible within a work of architecture, because it communicated the freestanding nature of buildings in a celebration of their physical presence. Christiansen designed to reveal the force-resisting capability of a building, stating that "visually apparent structural clarity has a universal appeal" and that "in a building, this expresses itself in a natural and orderly flow of loads from the top of the building down through the building structure to the foundations."[4] A core part of Christiansen's talent was envisioning the unity of structural forms and their reciprocal architectural space and being able to adapt these strong, stable shell structures to accommodate many different types of building functions.

THIN SHELL MODERNISM

Christiansen was certainly not alone in his admiration of shell structures. Modern architects and engineers used thin shell concrete for a variety of building types, with great variation in their design, throughout the twentieth century. Concrete shells became popular in France and Germany in the early twentieth century as a low-cost building option in times of material scarcity.[5] The use of thin shell concrete quickly spread throughout Europe, during the interwar period, before making its way to the Americas.[6]

In the United States, thin shell concrete played a vital role in midcentury modern architecture. In the postwar period, a significant strand of the modernist project looked to more expressive forms as a release from the rectilinear boxes of the International Style, demanding more alignment with modern structural engineers—just as Christiansen was beginning his design career. In 1954, Sigfried Giedion observed that the new era of modern architecture sought to solve the "vaulting problem," where "structural engineers [could] provide the stimulus to push the architect into new spatial adventures."[7] Eero Saarinen's Kresge Auditorium at the Massachusetts Institute of Technology (MIT) (1953–55), done in collaboration with engineer Charles S. Whitney, ignited significant debate in both architectural and engineering communities on the appropriate forms for shell structures.[8] Globally, Jørn Utzon's stunning yet controversial Sydney Opera House (1959–73) took these discussions even further and was only made possible by Ove Arup's groundbreaking work in structural engineering.[9]

Christiansen was at the forefront of the wave of architectural interest in concrete shell structures in the United States. He completed his first shell project—Seattle's Green Lake Pool—in 1954. By 1960, when architectural critic Ada Louise Huxtable stated that concrete shells were "undeniably today's most significant architectural adventure," Christiansen had already designed thirty separate shell structures.[10] Louis Kahn's Kimbell Art Museum (1968–72), in Fort Worth, Texas, used post-tensioned concrete shells, designed by structural engineer August Komendant, to organize the space and shape the light within a modern art gallery.[11] Christiansen had used a similar post-tensioning technique in the design of his shells as early as 1958.[12] While other monuments of modern architecture are certainly more well known, Christiansen's shell work was equally expressive and innovative.

Shell structures became a significant part of the modern architectural legacy of the postwar, midcentury period. The broader history of a "thin shell modernism" charts both an evolution within architectural modernism—more open to expressive form, but also in a deeper, more complicated, intertwined relationship with structural engineers—and, as shells began to fall from favor, a transition to a postmodern condition. In this light, Christiansen's work can certainly be seen as a significant instance of a thin shell modernism.

While many designers experimented with thin shell concrete in the 1950s and 1960s, or used it on individual projects, Christiansen passionately dedicated himself to this specific structural medium throughout his long career. Even though he designed with other types of structures at times, with considerable creativity, he always returned to thin shell concrete, regardless of its popularity among other designers or the general public. When he began in the 1950s, thin shell concrete was a cutting-edge, almost avant-garde approach to structural engineering design problems and a driver of modern architectural form. But by the end of his career in the 1990s, thin shell concrete had largely fallen out of use, replaced by other systems and architectural priorities. After its demolition, the Kingdome was replaced by stadiums with open-air, long-span steel trusses. Still, even as designers across the globe turned to other structural systems, Christiansen remained steadfast in his commitment to thin shell concrete.

This dedication reveals that, for Christiansen, thin shell concrete was more than a passing trend; it was the embodiment of his deeply personal philosophy toward architecture and engineering as a unified whole. The specifics of thin shell concrete, both the practicalities of its construction and the poetics of its potential forms, were inextricably linked to his design values and the position he maintained on what it meant to build.

NORTHWEST MODERNISM

As much as Christiansen was connected to the global legacy of thin shell concrete design, he was equally rooted in the design culture of the Pacific Northwest. The character and the quality of his shell designs are at home among the other contemporary architectural works known as Northwest Modernism.

Christiansen found that his key to success in the design of shell structures was in the use of a precise, reusable formwork system—one that could both lower costs and set the aesthetic profile of his work. As David Billington once wrote, the forms of Christiansen's shells "did not come from mathematical analysis but from simple structural and constructional ideas."[13] By using a single form to cast concrete on top of multiple times, Christiansen established a pattern of repetition and used the combination of multiple elements to make up a structural whole.

This compositional approach also mirrored the qualities of architectural modernism in the Pacific Northwest. Before Christiansen arrived in 1952, modern architects in the region had already emphasized the repetition of exposed structural elements, like evenly spaced wood or steel beams, in their architectural compositions. University of Washington professor and architect David Miller commented: "The Northwest School had a very strong reputation in the '40s, '50s and '60s. Characteristics include wood construction, repetition of columns and beams throughout the primary structure, transparency in the building by using glass."[14] In the documentary film *Modern Views: A Conversation on Northwest Modern Architecture*, architects of the time, including Ralph Anderson, and architectural historian Grant Hildebrand remarked on the strong interest in low-cost construction methods as a driver of their architecture.[15]

Christiansen worked as a consulting structural engineer, hired by his architectural clients. He contributed his interest and expertise in thin shell concrete to this existing design context. He not only found resonance in this approach but became a key contributor to its trajectory. As his career progressed, he often took a lead role in determining the structural form and spatial condition of certain projects, contributing creative, technical, and practical insight. Christiansen worked closely with significant architects such as Paul Kirk, Fred Bassetti, the early partners of Naramore, Bain, Brady & Johanson (NBBJ), and Minoru Yamasaki, contributing his own perspective in their collaborative work. With an ability to communicate across professional boundaries and appreciate the demands of each, Christiansen became a preferred collaborator of both architects and contractors.

Yet Christiansen also brought something else to Northwest Modernism: an ambition to work at a grand scale. As his confidence in structural design grew, Christiansen embraced projects of record-breaking span and size. Eager to test his architectural and engineering design ability, he sought out opportunities to create awe-inspiring buildings and spaces that could become modern civic monuments. These ambitions tie directly to his love of the outdoors and the towering mountains that surround life in the Pacific Northwest. An avid mountaineer, Christiansen summited over one hundred peaks in the Olympic Mountains, reached the top of Mount Rainier several times, and, later in life, traveled the globe in search of new mountains to climb. This adventurous spirit, and an ambition to challenge himself, is equally present in his designs—giving his work even more distinction within the canon of Northwest Modernism.

This book looks closely at these converging areas of emphasis—Christiansen's individual creativity, his connections to a global thin shell community, and the regional impact of his work. It shows that Christiansen's thin shell concrete work is creative yet technical, collaborative yet independent, regionally located yet globally significant. Understanding these dichotomies enriches and deepens an appreciation of Christiansen's role as a significant shell designer in a particular location, at a particular time. His work defies simple categorization and opens up many different realms of inquiry—such as the presence of a more technical version of global modernism than is typically recognized or the deep but infrequently studied relationship between architecture and engineering.

The book is organized chronologically, highlighting the different transitions and eras of Christiansen's career. Chapter 1 traces his youth in the suburbs of Chicago, his education, and his early career. Christiansen's desire for a life of adventure, formed in these early years, was essential to shaping his future work. Chapter 2 describes Christiansen's arrival in Seattle in 1952 and his early barrel-vaulted shell designs. Chapter 3 addresses Christiansen's adoption of the hyperbolic paraboloid as a driver of his thin shell forms and the importance of repetitive formwork in the economics of shell construction. Christiansen's work at the 1962 Seattle World's Fair—an important site and moment in the growth of the Pacific Northwest in which Christiansen played a prominent role—is the focus of chapter 4. Chapter 5 shows Christiansen's transitioning design approach, in which he adapted his thin shell forms to address changing architectural demands. Chapter 6 discusses the Seattle Kingdome—a complex project, deeply tied to the changing city. Chapter 7 describes Christiansen's later career and the demolition of the Kingdome. The conclusion brings Christiansen's work into the present day, speculating on his legacy and the possible future of thin shell concrete.

Throughout his career, Christiansen displayed a remarkable talent for developing innovative building forms that were expressive and beautiful, but also economical to build and materially efficient. His work had a profound impact on the people and culture of the Pacific Northwest.

CHAPTER 1

BEGINNINGS

Education in the American Midwest, 1927–1952

T HOUGH THE CITY of Seattle and the Pacific Northwest figure prominently in Jack Christiansen's career, his early life and education were rooted in Chicago and the American Midwest. The culture of Chicago and its surrounding neighborhoods—industrial, constructive, and diverse—became the stage of Christiansen's youth, instilling a peaceful yet practical understanding of the world. His love of reading introduced him to new, faraway places and inspired his future outdoor explorations. His education at the University of Illinois at Urbana-Champaign exposed him to the Midwest legacy of building innovation, kindling his love of architecture. While Christiansen's early experiences instilled values and perspectives he carried his entire life, they also set the stage for his departure to Seattle.

CHRISTIANSEN'S YOUTH

John Valdemar (commonly called Jack) Christiansen was born on September 28, 1927, in Chicago. He lived with his parents in a two-story duplex in the West Ridge neighborhood on Chicago's north side, near Grandville Avenue. Once composed primarily of bungalows, the area had densified during the 1920s with the construction of closely spaced, multifamily buildings. The duplex was owned by Christiansen's maternal grandparents, who lived upstairs, while Jack and his parents lived on the lower floor. Close to neighborhood parks and schools, Jack Christiansen's childhood home was comfortable and safe.

Jack's mother, Louise Linderoth Christiansen (1903–93), was the daughter of Niles and Cecelia Linderoth. The couple emigrated from Sweden in 1894, and Niles established a private dental practice in north Chicago. Louise grew up in Chicago and majored in mathematics at Northwestern University, graduating in 1926—often the only woman in her classes. After college, she worked at the Continental and Commercial National Bank in Chicago, but left the position when she married.[1]

Jack's father, Christian Valdemar Christiansen (1902–65), called C. V. or Wally, worked for the Bowman Dairy Company, just as his father had.[2] C. V. had attended the University of Illinois in dairy science and graduated in 1924. By the 1920s, Bowman had become the largest dairy in the Chicago area. Christiansen described his father's position as "dairy technologist," an emerging profession that used scientific processes to develop new dairy products for the consumer industry. Both scientific and commercial, his profession is listed in the 1930 US Census as "bacteriologist" and later, in 1940, as "chemist" for "retail, milk dairy." Christiansen recalled his father coming home from work and sharing samples of new types of ice cream or milkshakes that were being tested for consumer markets. C. V. Christiansen was successful enough in his career to survive the Great Depression of 1929 without significant economic difficulty. Jack Christiansen learned later that his father's salary had been reduced for one month during the Depression, only to be restored the next month. His father steadily advanced within the Bowman Dairy Company and eventually became the director of laboratories.[3]

Christiansen's father was likable and easygoing, making time after work to play with Jack and his younger sister, Joan (born 1935). They enjoyed some sports together, yet his father was also hampered by an undiagnosed heart condition and high blood pressure, and he often came home from work exhausted.[4] As another consequence, family vacations were casual affairs— like relaxing road trips to the rivers and lakes of northern Wisconsin. While enjoyable, these trips often left young Christiansen wanting more adventure and excitement.

With successful, supportive parents, Christiansen pursued a variety of interests. From a very young age, Christiansen took to drawing. As a child, he routinely drew family members' portraits—his mother and father, his nearby aunts and uncles. Recognizing his interest and ability, his parents sent him to take classes at the Art Institute of Chicago, one of the largest cultural institutions in the city.[5] Beginning at age eight, Christiansen would walk to the Granville Avenue station and take the Red Line train downtown—all by himself—a bit of freedom he greatly enjoyed. Christiansen recalled running through the halls of the Art Institute, surrounded by world-class artwork,

Jack Christiansen and sister Joan, 1936

before sitting down for instruction in drawing and painting.

Christiansen showed promise. One summer, a landscape he produced—a simple painting of a rural house, fence, and road—was selected by the Art Institute to travel the country as part of an exhibit of children's artwork. Christiansen recalled a deep sense of pride in his artistic creations, even at a young age. He would continue to draw and sketch throughout his personal and professional life.

In 1937, the Christiansen family moved to the suburb of Oak Park, into a large, two-story house at 802 Woodbine Avenue.[6] There were wide streets, expansive lawns, and large spaces between the suburban homes—it was a quiet, peaceful neighborhood. Christiansen enjoyed riding his bike along the tree-lined streets, and only occasionally did he notice the unusual, prairie-style homes of Frank Lloyd Wright dotted throughout the neighborhood.[7] Christiansen was, instead, more interested in taking the elevated train into Chicago for baseball and football games on the weekends.

Within his stable suburban life, Christiansen began to seek out more excitement, often within the pages of books. As a young boy, Christiansen became an avid reader and devoured any book he could get from school or the local library. He most enjoyed books with travel, intrigue, and adventure, like the Hardy Boys series. These books provided a window into alternative worlds outside of his own, introducing him to different places, people, and ideas.

Christiansen was particularly taken with *Lost Horizon*—the famous post–World War I adventure novel by James Hilton.[8] In the book, a war veteran exhausted by the burden of combat finds the utopian, mountain monastery of Shangri-La—a place of meditation and peace that fosters exceptionally long life. Young Christiansen was most taken by the descriptions of the stunning high-altitude location of the mountain retreat. In *Lost Horizon*, Hilton described the mountain landscape of the Himalayas in poetic detail: "Far away, at the very limit of distance, lay range upon range of snow-peaks, festooned with glaciers, and floating, in appearance, upon vast levels of cloud. . . . There was something raw and monstrous about those uncompromising ice-cliffs, and a certain sublime impertinence in approaching them thus."[9]

The mysterious mountains were natural, dangerous, and yet strikingly beautiful. Through Hilton's words, Christiansen could visualize this mystic mountain landscape, full of unexplored valleys and unknown peaks. The natural environment could create a poetic, transcendent experience. This was a world entirely unlike anything he had seen in the flatlands and rolling hills of the Midwest, and it captivated Christiansen's imagination.

Christiansen began to seek out images of real mountains in magazines such as *National Geographic*. He studied world atlases and maps, locating high peaks around the globe and devouring any information he could find on climbing adventures. He discovered a 1939 *Life* magazine article that began to describe for him a place like the one he had read about. Under the heading "America's Future," the article described the Pacific Northwest as a place of rugged abundance, where "the land is rich in nature's goods" and "irrigation makes the Northwest land bloom."[10] The article's accompanying photographs showed large expanses of land in Idaho, Oregon, and Washington that were being transformed through industry and infrastructure—such as the aluminum processing now made possible because of hydroelectric dams, including the Bonneville Dam and the soon-to-be-completed Grand Coulee Dam. The photographs showed jagged mountain landscapes and staggeringly tall stacks of wood boards in a Seattle lumberyard.

But unlike in *Lost Horizon*, this was a real place, one Christiansen could imagine visiting. Describing the very place he longed to find, the article proclaimed: "The look of the land bears out the Northwest's frontier promise. Behind the cities of the coast lie mighty reaches of forest, mountain, valley and river where you may go for miles and see only a thread of railroad track or a lonely settler's clearing as evidence of man's presence on the giant earth."[11] The Pacific Northwest had captivated Christiansen's spirit, and by age twelve, he began to envision a life for himself—one full of exploration and adventure—out West.

In fall 1941, Christiansen started his freshman year at Oak Park and River Forest High School. World War II had begun in Europe in September 1939, and the escalating war had a significant impact on his high school experience. After December 1941, when the United States declared war on Japan following the bombing of Pearl Harbor, Christiansen began to see older neighbors and friends heading off to war, and it became clear to Christiansen that he would be enlisting in the military upon graduation. In preparation for a technical occupation during the war, Christiansen focused on foundational subjects like science and math in his coursework. He did not take formal art classes in high school, although he continued to draw and sketch on his own. Like most boys at the time, he played sports in season: football in the fall, basketball in the winter, and baseball in the spring. Growing rapidly to over 6 feet tall, he particularly enjoyed basketball.

Christiansen also held many jobs over the summer as a youth—both before and during high school. With a shortage of workers due to men's obligatory military service, Christiansen had no trouble finding a job.[12] For several summers during high school, Christiansen worked in construction— his first real engagement with any type of building activity. He spent the entire summers of 1942 and 1943 working as a hod carrier—hauling bricks in wheelbarrows and mixing mortar for the masons at different places on a job site. Through the backbreaking labor, Christiansen began to see how buildings came together, all the while learning basic skills from the rough-and-tumble bricklayers. Christiansen kept his eyes open on each job, not realizing that this sort of activity would play a large role in his life.

By the time of his senior year (1944–45), Christiansen was playing junior varsity football and basketball. He was also a member of the Burke public speaking and leadership group and the Newton Club—an organization that "broadened knowledge of chemistry through talks, demonstrations, movies and trips."[13] The year Christiansen graduated, the school yearbook devoted several pages to students who had left school and were currently serving overseas, and Christiansen anticipated enlisting immediately after graduation. In early 1945, his parents took him to a local recruiter and enrolled him in a Navy radar technician program. Christiansen graduated from Oak Park and River Forest High School in 1945 and did not work at all that summer in anticipation of joining the military in the fall.

Jack Christiansen, relaxing in summer 1945

But global events drastically shifted Christiansen's path. World War II in Europe ended abruptly, with the German surrender on May 7, 1945. The United States' detonation of the atomic bomb over Hiroshima on August 6, 1945, effectively concluded conflict in the Pacific theater. At the end of the summer, the navy ended the radar technician program that Christiansen was set to enter. With the obligation to serve diminishing, rather than sign up for general enlistment, Christiansen and his parents decided—relatively quickly—that he should begin his collegiate education.

The only school Christiansen considered was the University of Illinois, in Urbana-Champaign, the state's flagship institution located roughly 150 miles south of Oak Park. The university was founded in 1867 as the Illinois Industrial University, a land-grant institution intended to provide industrial education to the public and foster the growing agricultural and technical needs of the state. John Milton Gregory and other early founders called the University of Illinois the "West Point for the working world," referencing the high-quality engineering education offered at West Point Military Academy.[14]

Christiansen did not have a specific career in mind as the fall semester approached. He felt no pressure to continue the family tradition in the dairy industry, nor did it hold much appeal to him. He remained interested in drawing, but also enjoyed his summers of construction and his science classes in high school. Searching for a specific path, Christiansen and his parents met with a high school guidance counselor, who gave him a career survey—a simple questionnaire that attempted to match students' interests with professions and degree programs. Much to their surprise, the survey fortuitously pointed to the major of architectural engineering—a subject completely unknown to them. Christiansen recalled asking the counselor, "What's that?," and only then did he begin to learn about the drawing, calculation, and design that went into the production of buildings.

ARCHITECTURAL ENGINEERING

Fortunately for Christiansen, the University of Illinois was one of only a handful of schools in the country to offer a degree in architectural engineering—a unique curriculum that blended education in architecture and engineering.[15] Started in 1890, the program at Illinois was the first in the nation. As envisioned by Dean Nathan Clifford Ricker (1843–1924), the program was intended to address the increasingly technical nature of architecture and construction in the United States.[16] In the late nineteenth and early twentieth centuries, the steel-framed, fireproofed skyscraper buildings of Chicago were ushering in a new paradigm of design and construction. Ricker foresaw the need for a hybrid professional with "an education in the field of architecture and the technical abilities to determine the structural design needed."[17] This program was providing an education not for structural engineers (a major offered through the College of Engineering) but rather for architects who sought advanced knowledge of the new types of structural systems that were redefining architecture and the technical basis to incorporate them fully in architectural designs. The program was meant to produce graduates who could work as architects, structural drafters, supervisors of construction, or similar positions involved in the production of buildings. The program also provided a foundation for future graduate work in structural design for those whose interests focused most strongly on engineering—a path that Christiansen would follow.

Ricker's training of architectural engineers at the University of Illinois was an attempt, in some ways, to reverse the nineteenth-century split between architects and engineers.[18] At the time that Ricker founded the architectural engineering degree program, schools of architecture across the country were largely following the French tradition of the École des Beaux-Arts. This tradition was rooted in the aesthetic history of architecture and emphasized careful drafting and artistic composition of building elements. Architecture was an extension of the fine arts and viewed more as a cultural than a technical endeavor. By the 1930s, institutions, including Columbia and Harvard, began to adopt new pedagogical models, based on the avant-garde German Bauhaus, leading to a nationwide transition to curricula aligned to the technology-centered Modern Movement.[19]

Conversely, schools of engineering in the United States had begun as military institutions (West Point as an example), modeled on the French École Polytechnique. The schools were intended to support the growing need for transportation infrastructure and industrial production through the teaching of scientific principles and technical skills.[20] Engineers from these schools were central in developing America's canal and railroad networks, but paid less attention to buildings. By the 1940s, engineering schools were becoming increasingly scientific, with more focus on controlled methods of research than on the practicalities of building and construction.[21] In addition to increased federal research investment in highways in the 1940s, engineers began to play an expanded role in American industry and business.

Ricker's curriculum occupied a space in between these architectural and engineering trajectories, maintaining his intent to give architects a thorough understanding of building construction.[22] Rather that follow French models, Ricker had visited the Baukademie in Berlin and was inspired by the merging of building science and liberal arts in the German curriculum model.[23] He also decided that an understanding of history was a central component of contemporary design.[24] As a result, the curriculum at the University of Illinois consistently explored a cultural yet technological American modernism. The arrival of the German Ludwig Mies van der Rohe in Chicago in 1938 was a seminal moment in the next wave of Chicago architecture. Success of this avant-garde approach to architecture depended greatly on the existing building tradition in Chicago.[25] Likewise at the University of Illinois, the architectural curriculum changed gradually with the influence of European Modernism in the 1940s, rather than undergo the dramatic adjustments that other schools did.[26]

This program played a vital role in setting Jack Christiansen's professional path. At the University of Illinois, he found a program where he could continue his interest in drawing but also learn state-of-the-art technologies and utilize his engineering sensibilities. He would have significant influence from the architectural culture of both Chicago and Germany. At other institutions, Christiansen would have had to choose either an engineering or an architectural education, but at the University of Illinois he could explore both together.

When Christiansen entered the University of Illinois in fall 1945, the program was led by Loring H. Provine, a practicing architectural engineer.[27] Officially, the architecture department at the University of Illinois offered two tracks for a bachelor of science degree—the general option (architecture) and the construction option (architectural engineering), which Christiansen selected. As described in the 1945–46 course catalog: "The *construction option* (architectural engineering) offers a major study in building design, a thorough training in all forms of building construction, and emphasizes the structural and mechanical aspects of architecture. As the curriculum includes two years of architectural design, freehand drawing, and the history of architecture, the student who is primarily interested in construction can acquire a considerable knowledge of the artistic and utilitarian phases of planning."[28]

This comprehensive, inclusive coursework instilled a deep sense of the connection between the architectural, engineering, and construction professions. During the first year of instruction, Christiansen's coursework included architectural design studio, freehand drawing, algebra, trigonometry, and analytic geometry—a curriculum shared by all architecture and architectural engineering students. The drawing courses were the most challenging for Christiansen, going far beyond anything he had tried before. He learned drafting techniques, as well as proportioning, perspective, and shading. He recalled: "I showed up to [the] University of Illinois and they put me to work right away making ink wash drawings of historical architecture. If you've ever done ink wash drawings, you know that if [you] accidently drop a little bit of ink someplace, you start over, so it was good training, because I had not taken drawing or anything in high school."[29]

The drawing of historical works of architecture was an important part of Ricker's pedagogical vision and central to Christiansen's design education. This experience gave Christiansen a profound appreciation for historical buildings and a personal connection to architecture through drawing. He began to grasp how drawing was a process of designing and a means to communicate complex building ideas difficult to describe with words. Christiansen's ability to draw would later distinguish him from other engineers and become a vital part of his future work.

Christiansen enjoyed other college activities, but not many. He joined a fraternity (Beta Theta Pi) and, during his freshman year, decided to try out for the basketball team. Soon, however, with a demanding course load, he was forced to make decisions about his priorities on campus. His architectural design studio courses, offered in the afternoon, often ran so long that they began to conflict with basketball practice. One afternoon, Christiansen approached his studio instructor about leaving early, only to receive a dramatic response.[30] He recalled: "She hit the ceiling. She became very animated and asked me: 'What are you doing? You're throwing your life away. For basketball?!' So, that was the end of basketball for me."[31]

From then on, Christiansen focused less on sports and more on his education. Despite the outburst, his studio professor saw talent in Christiansen's work and gave him strong marks. In other courses, with class sizes of fifty to sixty people, Christiansen performed well and was recognized for getting the highest grade in several classes.[32]

During the second year, all architecture students took the first of four semesters of History of Architecture, keeping with Ricker's original intent.[33] At the time of Christiansen's arrival, Associate Professor Thomas E. O'Donnell taught the architectural history courses, described in the catalog as "an analysis of structural space and form," a telling of history that linked the buildings of the past to contemporary design problems. Addressing various periods and geographic locations (Egypt, Persia, Greece, Rome, the European Renaissance, the Americas, and modern Europe), the history sequence was based on study of "environmental influences and scientific achievement" as a driver of architecture.[34]

The historical perspective that Christiansen gained through these courses was central to his thinking. As a result, he understood architecture and structural engineering as interrelated, mutually dependent professions. Later in his life, Christiansen often claimed that "the history of architecture . . . *is* the history of structure."[35] He understood the evolution of structural forms—like the post-and-beam construction of Greek temples, Roman arches and domes, and the buttresses of Gothic cathedrals—as central to the evolution of architecture.

Christiansen later reflected on the importance of history as something that distinguished his education from that of other engineers: "Engineering education has never placed an emphasis on the history of architecture, which is too bad, as today's graduating engineers lose out in not understanding the work of their forebearers."[36] Through his education, Christiansen understood that he was entering a profession with a long history.

INTRODUCTION TO CREATIVE ENGINEERS

The architectural history courses introduced Christiansen to contemporary ideas as well. The final history course described the "gradual change from traditional to modern methods of construction," providing a history of modern architecture up to the present day (for Christiansen, the late 1940s). In this course, Christiansen encountered the work of Eugène Freyssinet (1879–1962) in France, Eduardo Torroja (1899–1961) in Spain, and Pier Luigi Nervi (1891–1979) in Italy—all pioneering designers of thin shell concrete structures. Christiansen recalled seeing Freyssinet's airship hangar at Orly (1921), Torroja's Zarzuela Hippodrome (1935), and Nervi's airplane hangars at Orvieto (1935–38) and marveling at the graceful structural forms. Each of these designers transformed reinforced concrete into lightweight, long-spanning structures through their own signature method and style. Through their creative engineering designs, Christiansen began to understand the complexity and possibilities of thin shell concrete. These designers, and their diverse approaches, directly inspired Christiansen's own pursuit of thin shell concrete as a design medium.

Christiansen's textbooks included *The Architecture of Bridges*, published by the Museum of Modern Art in 1949. This text, written by Elizabeth Bauer Mock, was later credited by David Billington as the "first book on bridges to offer a modern viewpoint."[37] Mock's "conviction that a fine bridge is also fine—and modern—architecture" was offered as "the basis of this book and its only justification."[38] The book showed bridges throughout history, such as Roman stone bridges and aqueducts (ca. 100), the Pont Neuf in Paris (1607), the Dolau-herian Bridge over the Towy near Llandovery, Wales (1750s), and several modern examples. The text described the careers of François

Christiansen's personal copy of *The Architecture of Bridges*

Hennebique, Freyssinet, and Robert Maillart (1872–1940), showing their bridge designs of varying types (concrete arch, prestressed precast concrete, and suspension bridges). Christiansen took particular notice of Maillart's Schwandbach Bridge (1933), Salgina Bridge (three-hinged arch, 1930), Thur Bridge (1933), and Arve Bridge (1937). Christiansen recalled seeing pictures of these bridges as a student: "[Maillart] at the turn of century did these fantastic concrete bridges over marvelous mountain gorges in Switzerland. And I looked around at the world, and I said, 'I don't see anything like this being done, this is wonderful work.' I loved it."[39]

Christiansen appreciated the forms of the gracefully designed bridges and the stunning natural peaks of the Swiss Alps. These images triggered both his sense of adventure and his creative spirit—reminding him of the descriptions from *Lost Horizon* and the images of the Pacific Northwest in *Life* magazine. Inspired by Maillart's work, Christiansen began to realize how, as a designer, his desire for a life of adventure and his interest in structure could come together.

Mock's book linked Christiansen's love of bridges, history, and travel all together. Christiansen would continue to be inspired by these bridges throughout his life. He would travel to visit these sites for himself and take his own photographs—as near reproductions of the images in the book.

As part of his architectural curriculum, Christiansen read *Frank Lloyd Wright on Architecture*, a collection of writings and lectures from 1894 to 1940.[40] Even though Christiansen had grown up in Oak Park, he had little exposure to Wright's work or writings prior to his university education. Christiansen was drawn to the principles embedded in Wright's writing, referencing Wright's built work in a separate book.[41] Within the curriculum text, Christiansen marked several passages that he found significant. These ranged from short phrases such as "I would much rather build than write about building" to longer passages on the deeper relationship between architecture, nature, and structure.

Christiansen also saw in Wright's work a connection to his love for poet Walt Whitman (1819–92). As a child, Christiansen had read Whitman's

Robert Maillart's bridge over the Arve River, Vessy, Switzerland, completed 1937

poems, and he kept a 1933 copy of *Leaves of Grass* with him at the university. Within the pages of the Wright book, Christiansen inserted his own handwritten transcriptions of poems by Whitman—portions of "A Song of the Rolling Earth" and "Song of the Open Road." An insertion from 'To the Sayers of Words,' reads:

> Say on . . . ! Dig, model, pile up the words of the earth. Work on—(it
> is materials you must bring not breaths) work on age after age—
> nothing is to be lost; it may have to wait long, but it will certainly
> come into use; when the materials are all prepared, the architects
> shall appear. <u>I swear to you the architects shall appear without fail!</u> I
> announce them and I lead them. I swear to you they will understand
> you and justify you; I swear to you the greatest among them shall
> be he who best knows you, and endorses all, and is faithful to all; I
> swear to you, he and the rest shall not forget you—they shall perceive
> that you are not an iota less than they; I swear to you, you shall be
> glorified in them.
>
> —WALT WHITMAN

Christiansen underlined the passage concerning the emergence of architects. Driven by his love of reading, he embraced the poetic side of architecture and its connection to the natural world, through a fusion of Whitman and Wright. These passages helped Christiansen develop confidence in his emerging abilities as an architectural engineer, while forging a philosophical connection between architecture, nature, and poetry. Perhaps most telling of his future design direction, as a synthesis of material and structure, Christiansen specifically highlighted one of Wright's definitions of architecture: "Architecture is the scientific art of making structure express ideas."[42]

ENGINEERING TO THE FORE

Only after his first two years of a primarily architectural education did Christiansen begin focusing on structural engineering coursework. The third and fourth years of architectural engineering education brought a split from the students who chose architecture as a general option. With no more studio classes, Christiansen faced more detailed engineering topics of study, such as the technology of materials, structural elements, the theory of structural design, and advanced structures. The only material-specific, structural design class taught in the curriculum was on reinforced concrete (not steel or wood)—offered as two semesters of theory and design.[43]

During these two years, Christiansen had several courses with Professor Newlin Morgan (1888–1958), including the first class he took in reinforced concrete design.[44] Morgan had graduated from the University of Colorado in 1910 with a bachelor of science in civil engineering. He had worked professionally as a structural drafter with the American Bridge Company and then as a structural designer with the Chicago, Milwaukee and St. Paul Railway Company until 1918.[45] By 1924, Morgan had joined the faculty of the Department of Architecture and had begun working with the noted engineering professor Hardy Cross (1885–1959). Leonard K. Eaton, in his biography of Cross, claimed that Cross was "the outstanding American of his time" in the field of engineering—a field dominated by Europeans.[46] Cross promoted a holistic view of structural engineering: "Engineers are not . . . primarily scientists. If they must be classified, they may be considered more humanists than scientists. Those who devote their life to engineering are likely to find themselves in contact with almost every phase of human activity."[47] This understanding of engineering had a great effect on Christiansen.

In 1932, Cross and Morgan coauthored the notable textbook *Continuous Frames of Reinforced Concrete*.[48] The textbook presented a new method of analyzing indeterminate beams and frames—engineering problems that could not be solved directly using algebra. Called the "Hardy Cross method for moment distribution," the method showed how simple calculations could be used to successively approximate a solution for the bending moments in a beam. By gradually approaching the correct solution, getting closer and closer with each calculation, Cross and Morgan eliminated the need for a longer, more tedious solution.[49] They emphasized that, to the practicing engineer, a close, approximate solution was nearly as useful as an exact solution—releasing the structural designer from the burden of impractical precision.

Morgan, whom Christiansen recalled as an effective lecturer and teacher, taught his courses in lengthy three-hour blocks. With Morgan's instruction, Christiansen learned how reinforced concrete was a technical, high-performance structural material—the same material that had been used in the elegant, thin shell designs from his history courses. Christiansen later commented: "I look at concrete as a very natural material, it's, you know, it's rock, man-made rock, and, it's available all over the world, . . . it's a very economical material, a very durable material, and it's a plastic material. In other words, it can be molded into any shape that you want."[50]

Under Morgan's teaching, Christiansen came to understand reinforced concrete as a multifaceted medium, with high performance but also a freedom of form unmatched by any other structural material. Having already captivated his imagination in the architectural history courses, reinforced concrete would also become the material at the center of Christiansen's structural engineering education.

During his fourth year, Christiansen met his future spouse—Suzanne Gertrude Hasselquist (1926–2010). She was a year older than he and had already graduated from Augustana College in Rock Island, Illinois, in 1948. Her great-grandfather, T. N. Hasselquist, was a prominent Lutheran minister and a founder of Augustana College. Suzanne (known as Sue) was a visual artist, then completing a second degree at the University of Illinois in art, and just beginning to work professionally as an illustrator. She was one of three artists who worked on the 1949 volume of *The Illio*, the University of Illinois annual yearbook.

Luckily for Christiansen, the art department was housed in the same college (the College of Fine and Applied Arts) as architectural engineering, and the two met on campus in 1949. In addition to sharing Christiansen's love of drawing, she enjoyed painting and pottery. She also shared his desire for a life of adventure and challenge and was interested in hiking, mountaineering, and travel. Together they established a lifelong partnership, full of art and nature.

Christiansen later called his four years at the University of Illinois a "life-changing experience." The unique combination of history of architecture and

Jack Christiansen during his time at the University of Illinois, ca. 1948

structural design courses (and instructors) had introduced him to the world of architecture and engineering and provided him with a career path that he was excited to pursue.

Embracing the two professions, Christiansen held student memberships in both the American Institute of Architects and the American Society of Civil Engineers (ASCE), as well as the Illinois Gargoyle Society, which promoted "scholastic achievement in architecture."[51] His undergraduate education gave him a deep grounding in the design and construction of buildings, as both a practical and a creative act. He recalled: "I came out of those four years really in love with architecture and engineering."[52] Christiansen received his bachelor of science in architectural engineering, with high honors on June 25, 1949.[53]

Even though he had completed his degree, Christiansen had only encountered thin shell concrete within the context of a history course, not through a structural design course. Christiansen found no courses offered at the University of Illinois in the design of shell structures. In order to pursue his growing interest, Christiansen determined that he needed more education in structural engineering, and Professor Morgan agreed.

> I remember going to talk to Newlin Morgan, my professor [at] Illinois in the last few weeks of school, and I went to him and I said . . . I got to learn about these shell structures I've been reading about, these slender bridges, and he agreed that I needed more education along those technical lines, and he told me about a professor who had just recently come to Northwestern University in Evanston, Illinois, who was a protégé of Hardy Cross at Yale, and he just got a doctorate and he was teaching there, so I enrolled in Northwestern University.[54]

The professor was Frank Baron (1914–94). Baron graduated from the University of Illinois in structural engineering, studying under Hardy Cross (before he moved to Yale), and received his doctor of science at Harvard in 1941. Baron taught alongside Cross at Yale before going to Northwestern in 1946.[55] Along with the other professors at Northwestern, Baron would give Christiansen the advanced technical knowledge in structural analysis he would need to begin a career in thin shell concrete design.[56]

In fall 1949, Christiansen enrolled in the master of science program in structural engineering at Northwestern University. His wife-to-be, Sue

Hasselquist, had recently accepted a job as an illustrator for a publishing house in Chicago. Christiansen was able to find an apartment in Oak Park, commute easily by train to Northwestern (in Evanston, just north of Chicago), and still be close to Sue. The two were married on March 12, 1950.

Christiansen now faced a more typical engineering curriculum of rigorous, scientific study of structural materials and their behavior in Northwestern's Department of Civil Engineering. One of Christiansen's classes was Advanced Mechanics of Materials—the "advanced applications of the basic principles of design based on analytical and experimental studies of theories of failure, plastic range stresses and energy loading." This included difficult engineering problems of that time, such as the behavior of curved beams, plates, stiffened sheets, and structural frames. Another class addressed structural analysis: "the application of classical and recent procedures of analysis to problems of structures, including such structural types as building frames, arches, rigid frames, continuous trusses, and suspension bridges." The methods taught in this class included formalized algebraic procedure, successive approximations (such as the Hardy Cross method), and conservation of energy procedures. Another class focused on structural design: "the application of theories of structural action to the planning and design of structures and their elements correlated with data obtained from experiments and experience." Here Christiansen encountered materials such as steel, timber, concrete, and aluminum in the design of various structural types, to understand current engineering practices and the complementary code requirements.[57]

At Northwestern, Christiansen received his first exposure to the theory and behavior of structural shells. His textbooks included Stephen Timoshenko's *Theory of Plates and Shells* (1940), which described the general behavior of thin, curved structures in a theoretical manner.[58] The equations in the book could be used for designs of a variety of materials and functions—from ship hulls to thin-walled pressure tanks and locomotive engines, as well as structural slabs. This broad application of shell structures introduced Christiansen to the general stress equations for shells but stopped short of providing a design method for shells for buildings in reinforced concrete.

FIRST PROFESSIONAL EXPERIENCE

Christiansen graduated from Northwestern University in spring 1950 and began searching for employment in downtown Chicago. His training as an architectural and structural engineer prepared him well to work in the professional climate of Chicago. Paralleling the advancements in building technology, the city had developed a long tradition of architects and engineers working together in one firm.[59] Christiansen got his first job that year with the architectural firm Perkins & Will. Architects Lawrence B. Perkins and Philip Will Jr. had founded the firm in 1935 and had built a reputation designing primarily schools and residences. They had hired their first in-house structural engineer just three years earlier, in 1947, making Christiansen one of the earliest structural engineers in the office. In 1951, the firm established mechanical and electrical engineering departments as well.[60] Christiansen worked on the structural design of several projects, most notably the Keokuk High School in Keokuk, Iowa—the firm's first $1 million building and a recipient of an American Institute of Architects National Award.[61]

Christiansen enjoyed the work at Perkins & Will, but he soon discovered that he could increase his salary by 50 percent by moving to another firm. With a wife and a baby daughter (Janet, born in 1951), Christiansen left Perkins & Will after about a year to move to Shaw, Metz & Dolio. This firm was founded as a partnership in 1947, led by Alfred Shaw (architect), Carl A. Metz (structural engineer), and John Dolio (mechanical and electrical engineer). The firm had expanded rapidly, designing department stores, banks, residences, and apartment buildings—and provided all architectural and engineering services in-house. Once again, Christiansen was surrounded by a professional environment that reinforced the connection between architecture and engineering. He was designing reinforced concrete structures, integrated into functional buildings, but still saw no opportunity for thin shell concrete.

That would soon change. One day, in 1952, Christiansen saw Anton Tedesko (1903–94)—a pioneer of thin shell concrete in the United States—walk into the office of Shaw, Metz & Dolio. Tedesko, an Austrian-born engineer now practicing in the United States, had previously worked in Europe for the German engineering company Dyckerhoff & Widmann. This firm, led by engineers Franz Dischinger and Ulrich Finsterwalder, had developed a method of design and analysis for thin shell concrete structures, which came to be known as the Zeiss-Dywidag (Z-D) system.[62] With an approach rooted in a German scientific tradition, Dischinger and Finsterwalder developed exact mathematical solutions for the structural behavior of certain thin shell geometries.[63] In parallel, the firm developed a geodesic system of precisely machined rods and connections to use as formwork for spherical domes, known as the Zeiss network. The company used these solutions to develop buildable shell structures and successfully marketed these shells as a low-cost building option. In the 1920s and 1930s, the Z-D system was widely used for the construction of simple barrel-vaulted warehouses, factories, and

F. W. Woolworth Department Store, Chicago, designed by Shaw, Metz & Dolio, completed 1949

centralized domed markets throughout Europe. Signature, large-scale structures in Germany included the Zeiss Planetarium in Jena (1922) and the Leipzig Market Hall (1929).

After finding success in Europe, Dyckerhoff & Widmann looked to expand its business to the American market. With the United States in the midst of the Great Depression, the firm hoped the low-cost shell construction system could become a viable construction option. Dyckerhoff & Widmann established a partnership with the American engineering firm Roberts & Schaefer in Chicago, agreeing to license the Z-D system in exchange for a percentage of the contract price for each project. Tedesko, who had prior experience living in the United States, came to Chicago as a representative of Dyckerhoff & Widmann in 1932 for a two-year trial period.[64]

However, Tedesko knew that the German shell techniques would have to be significantly adapted to suit the American context—where the success would be judged entirely on construction cost efficiency in a drastically different labor market.[65] The American municipalities also did not allow the use of foreign products, thereby prohibiting the proprietary Zeiss network of formwork construction.

Through close collaboration with American contractors, Tedesko oversaw the development of new construction methods and the application of formwork systems that saved on labor costs but deviated from the mathematical precision of the German shells. Tedesko supervised the structural testing of these modified shell forms, ensuring that they could be as safe as their German counterparts.[66] Tedesko explained: "American shells are different from their foreign counterparts, with the designs emphasizing labor savings at the expense of additional material. Ingenious contractors contributed new ideas and new job applications, as the design of forming and the handling of labor were of much greater importance here than they were in Europe. These improvements in construction technique made possible the success of American shells, which were the product of resourceful designers who visualized the construction job as it would be built while the structure was still in the design stage."[67]

Tedesko also effectively negotiated the political and regulatory restrictions in the United States, allowing thin shell concrete to be accepted for large-span structures. In 1934, Tedesko oversaw the construction of the domed Hayden Planetarium in New York City, with a diameter of 81 feet and a thickness of 3 inches—one of the first, full-scale American thin shell concrete structures. Later, in 1936, Tedesko designed the barrel-vaulted Hershey Arena, in Hershey, Pennsylvania, a structure that spanned 225 feet with a thickness of only 3.5 inches.

Hayden Planetarium, New York City, with a thin shell concrete dome designed by Anton Tedesko, completed 1934

With the slow economy during the Depression, and later the severe shortage of steel during World War II, Tedesko's concrete thin shells gained in popularity. Into the 1940s, Tedesko designed a series of airplane hangars for the US government, as well as several warehouses and gymnasiums.[68] By 1950, Tedesko's efforts had resulted in several million square feet of American thin shell concrete and focused primarily on marketing shells to other design firms.

It was this role, as a marketer of shells, that brought Tedesko into the offices of Shaw, Metz & Dolio. As a young engineer in the firm, Christiansen did not immediately recognize Tedesko, nor did he have a complete understanding of Tedesko's work at the time. Tedesko asked to speak with Metz, the lead structural engineer in the office, hoping to convince him to consider a thin shell concrete structure on upcoming projects. Tedesko described the material and cost advantages of thin shell concrete construction and how suitable this form was for certain building types. Tedesko was offering complete structural engineering design services for a thin shell concrete structure that could suit a variety of project needs.[69]

Metz listened to Tedesko's sales pitch, but politely declined his services. After Tedesko left, Metz, still intrigued by thin shell concrete, came over to Christiansen's desk and asked if *he* could design a thin shell of concrete. Christiansen considered his education in the theory and mechanics of structural design. He recalled his fascination with the efficient, elegant forms of Nervi, Maillart, and others. He wanted to be able to say yes and begin to design these structures on his own. But Christiansen responded no. While he understood some of the theory of thin shell structures, he simply did not have the confidence or construction experience to execute a design of his own. The time-sensitive design process and the construction cost implications of each project were too great for Christiansen to take on at this stage in his early career.

But still, the question from Metz spurred Christiansen's interest, and he began to seriously ask himself how he would design a thin shell concrete structure. From that point on, thin shell concrete structures were no longer just images in history courses but real possibilities. He began to envision shells as structural solutions to the engineering problems he was addressing in his everyday design work. While he still needed to determine his own method of structural analysis, he was gaining more and more practical experience in conventional concrete construction. He was close, but not quite ready.

At the same time, a broader interest in thin shell concrete structures was expanding within the architectural and engineering community in the United States. In June 1950, the journal of the American Concrete Institute (ACI) published the article "Cost of Long-Span Concrete Shell Roofs" by American engineer Charles S. Whitney.[70] Whitney, as part of the firm Ammann & Whitney, would soon be tapped by Eero Saarinen to assist in the design of MIT's Kresge Auditorium—one of the first architecturally significant shell designs in the postwar United States. The ACI article described the economics of shell construction (including material and labor), advocating thin shell concrete as a cost-effective building technique, if carried out properly.

Whitney's article echoed much of the work done by Tedesko, in proving thin shell concrete's long-span capability. In December 1951, a relatively unknown Spanish-Mexican builder named Félix Candela contributed the article "Simple Concrete Shell Structures" to the ACI's journal.[71] Despite its title, Candela reported on his experiments with a variety of increasingly complex concrete shell forms that began to show the material economy and expressive potential of shell design. Christiansen read these articles, and his interest in shell construction grew even more.

Christiansen enjoyed working at Shaw, Metz & Dolio, and could envision a promising career there, but, ultimately, he wanted more adventure than Chicago had to offer. He had grown tired of his daily commute to the bustling downtown center, through the crowded trains and busy streets. Christiansen had never forgotten the mountain landscapes he had read about in his youth, and the idea of exploring the mountains for himself still captivated his imagination.

In early 1951, Robert Claus—a relative—passed through Chicago and stopped for dinner at Christiansen's parents' house in Oak Park. Jack and Sue came over to visit. Claus was the husband of Christiansen's cousin Janet and was driving cross-country from Detroit back to Sedro-Woolley in Washington State—a small town at the base of the North Cascade Mountains.[72] Over dinner, Claus shared stories of life in the Pacific Northwest, describing their farm and rural lifestyle in the foothills. He told of hunting and fishing trips by pack horse, deep into the mountainous wilderness. Christiansen was captivated.

The firsthand accounts of these dramatic, natural experiences reminded Christiansen of his childhood adventure books. The stories he heard about the Cascades, the Olympic Mountains, and Mount Rainier reminded Christiansen of his boyhood infatuation with the fictional mountains surrounding Shangri-La. The region's deep gorges recalled the setting for Robert Maillart's mountain bridges in Switzerland that he had admired while at the University of Illinois. Yet these snowcapped mountains and alpine lakes were within driving distance of a growing city—Seattle—where Christiansen could continue his career as a structural engineer. Christiansen remembered the *Life* magazine article that described the growing industries of the Northwest.

Sue was just as excited as he was. He later recalled: "Our motivation for moving to Seattle in 1952 was our desire for a balanced life of working in a city and discovering the natural wonders of the West."[73]

By the end of that dinner, much to the shock of their parents and extended family in the Midwest, the Christiansens had decided they would move their family and set up a new life in the Pacific Northwest.

With enough money to buy a used car, in the summer of 1952 they packed up all their belongings and started driving west. Christian-

Jack Christiansen preparing to drive to Seattle in 1952

sen had another relative—his aunt, Mary Christiansen Hulltein—who lived in Bremerton, on the Olympic Peninsula. She had offered the Christiansens a place to stay while they got settled in the Northwest, so they set out for Bremerton. Intent on seeing as much of the country as they could, they stopped at several National Parks along the way. When they arrived at Grand Teton National Park in Wyoming, Christiansen was overwhelmed by the size of the mountain range, claiming he had "never seen a real mountain before that day." As they arrived at the park, a group of mountain climbers were descending the range—and Christiansen stopped to talk to them. They told him about their trip, the struggle of mountain climbing, and the reward of reaching the top. Christiansen was excited about the prospect of climbing, the testing of one's self against the natural conditions, and was determined to try it himself in the Pacific Northwest.

The Christiansens took the northern route into the Puget Sound region (US Highway 2) across Stevens Pass and then detoured south through the mountains to Snoqualmie Pass (now Interstate 90). The rugged mountain roads were then only narrow tracks, winding around peaks and surrounded by boulders. As the road descended through the foothills, down to a glittering Lake Washington, Christiansen saw Seattle for the first time.

CHAPTER 2

FIRST FORMS

Design with Cylindrical, Barrel-Vaulted Shells, 1954–1958

U PON HIS ARRIVAL in Seattle, Christiansen discovered a rapidly expanding city that offered many opportunities to build. Finding work with the oldest structural engineering firm in the city, he began his practice in thin shell concrete—aided by an industry design manual, *Design of Cylindrical Concrete Shell Roofs*. Surrounded by a large need for all types of buildings, Christiansen designed barrel-vaulted, thin shell concrete structures for pools, warehouses, schools, churches, airplane hangars, and other building types. Within a few years, he began extending the structural and spatial capability of the cylindrical form beyond the specifications of the manual, demonstrating an emerging creative approach.

A POSTWAR SEATTLE

After World War II, the economy of the Pacific Northwest was booming—with several key industries supporting widespread growth, but none more important than the Boeing Aircraft Company. Boeing, founded in the Pacific Northwest in the early twentieth century, had been a primary supplier of aircraft during the war, employing thousands throughout the region. As defense spending once again increased with the dawn of the Cold War and the Korean War, Boeing reemerged as a major economic driver of the region. As Boeing modified and improved its airplanes, it placed constantly changing demands on the infrastructure and labor of the Northwest, requiring new, larger facilities and an increasingly technical labor force. The majority of

Seattle waterfront and skyline, 1952

engineers in the Northwest, regardless of specialization, maintained a connection to Boeing in one way or another.

Other industries were emerging as well. Timber companies in the region, such as Weyerhaeuser and the Douglas Fir Plywood Association, began moving beyond pure resource extraction and expanded their production of engineered wood products. Products like plywood, laminated veneer, and glued laminated (glulam) beams were first used in small-scale experimentations and wartime applications but soon found uses in the domestic construction industry, ushering in new building forms and new ways of building.

As a result of a healthy economy and these advances in industry, population was on the rise as well. Spurred by veterans returning from World War II and an abundance of new jobs, King County's population increased 45 percent between 1940 and 1950, and then 25 percent again before 1960. This growth placed new requirements on the region's infrastructure of roads, bridges, and residences. Due to the effects of the Great Depression and wartime austerity measures, the region had a large, unfulfilled demand for nearly all types of buildings, including schools, office buildings, residences, libraries, and churches. After decades of minimal construction activity, building projects began to flood the offices of architects, contractors, and structural engineers.

The region's professional community of architects, engineers, and builders was redefining itself as well. With plenty of work to go around, architects, who had formed strategic partnerships during the war years to stay afloat, often separated, and young architects (many only one or two years after graduation) opened their own firms. At the same time, architectural tastes had changed a great deal since the Great Depression, and an architectural modernism was sweeping through Puget Sound. Architects educated in the 1930s and 1940s had a distinctly different outlook from those educated in the 1920s. In the Puget Sound region, thin lines and flat roofs began to replace the ornament and detail of Art Deco and streamlined Art Moderne, showing the influence of the European Modernists (such as Le Corbusier and Mies van der Rohe) and the International Style.

Young Northwest architects now found themselves with an opportunity to design buildings in a Modern style, with a new freedom to try different conceptions of space and incorporate developing structural technologies. Architects began to use exposed structural steel frames and large expanses of glass in their buildings, creating a different type of architecture. While it was distinct from the building community Christiansen had encountered in Chicago, he was familiar with the necessary dialogue with structural engineers that this architecture demanded.

Structural engineers in the Puget Sound region were experiencing their own version of modernization. The effects of the Great Depression and World War II caused a significant decline in building activity. During this time, many engineers found work with Boeing or the naval shipyards in Bremerton. When a large earthquake hit western Washington in 1949, the structural engineers in the region came together to decide on new seismic requirements for building design—forming the Structural Engineers Association of Washington. New buildings would have to be stronger than ever before, but engineers were now equipped with better grades of steel, stronger concrete mixes, and new methods of calculation. These new skills and techniques would define the coming era of building construction and give form to emerging Modern architecture.

Flush with work, expanding firms began to rapidly hire young professionals, often fresh out of school—providing opportunities and responsibilities that far outpaced their experience. Both architecture and engineering firms relied heavily on recent graduates of the University of Washington, keeping close relationships with university professors. This rapid expansion infused the industry with a youthful exuberance and a professional culture open to exploring new ideas and trying new things. Jack Christiansen would have no trouble finding a job.

JOINING THE W. H. WITT COMPANY

In September 1952, Christiansen walked into the offices of the W. H. Witt Company. When Christiansen had announced he was coming to Seattle, Carl A. Metz had suggested the offices of the structural engineer Sigmund Ivarsson as a place to work.[1] Ivarsson was an older engineer in Seattle and was well known for his work with the local architect Paul Kirk (he would later be the engineer for the University of Washington Faculty Club, 1958–60). But Ivarsson was getting ready to retire and wasn't interested in hiring at the time. Ivarsson referred Christiansen to the W. H. Witt Company.

At the W. H. Witt Company, Christiansen had unknowingly found a firm that had a long history in the Northwest yet was no stranger to innovation and change. The W. H. Witt Company was, in fact, the only structural engineering firm in Seattle to survive the Great Depression, and it kept continuous operation through the wartime years. The company had been started in 1924 by William Henry Witt and made its name designing midrise concrete office buildings in Seattle, such as the Lloyd Building (1926) and the Vance Building (1927). However, after Witt's untimely death in a railroad crossing accident in 1928, two younger engineers—Harold Worthington (1902–94) and George Runciman (1891–1965), both graduates of the University of Washington—took over the leadership of the firm. In 1931, the W. H. Witt Company designed (arguably) the world's tallest reinforced concrete skyscraper in Seattle (the Seattle City Light Building), only to have construction halted due to the Great Depression.[2] In the 1930s, the company designed innovative concrete bridges throughout the region, working closely with Homer Hadley (designer of the first floating bridge on Lake Washington).

In the 1940s (during the war and just after), Worthington and Runciman designed state and federal buildings in Olympia and naval buildings on the Olympic Peninsula. They kept tight relationships with their architectural clients—most notably the firm of John W. Maloney and a newly formed architectural partnership named Naramore, Bain, Brady & Johanson (NBBJ). These two firms attracted a number of large federally and state funded projects in the late 1940s and early 1950s, with enough work to keep the W. H. Witt Company in business. During this time, the firm employed only three to four people, but, more importantly, it maintained a close bond with architects in the region. When the economy began to expand again, the W. H. Witt Company was well positioned.

In September 1952, Christiansen interviewed with Worthington, the managing partner. Christiansen presented a roll of drawings that he had produced while working at Perkins & Will in Chicago—a set that he had both designed and drafted. Worthington was impressed with Christiansen's range of capabilities and offered him a job on the spot.

At the firm, the small community of colleagues and friends helped Christiansen and his wife, Sue, get settled into life in the Pacific Northwest. Joe Jackson (1906–2001), an older engineer in the office and a resident of Bainbridge Island, invited Christiansen to come over to his place on the weekend. Jack and Sue were still living with his aunt in Bremerton and looking for a place of their own. One Saturday, they took the ferry from downtown Seattle to Bainbridge Island and were enthralled with the surrounding waters and mountains. Christiansen discovered that, viewed from the west side of Bainbridge Island, the Olympic Mountains seemed to rise directly from the waters of the sound into the clouds above. It was the setting he had dreamed about since he was a child and contained all the potential adventure he was hoping for. Shortly after that weekend, Christiansen connected with a real estate agent named Sam Clark and rented a small beach cottage with views of Seattle and the Cascade Mountains for $50 a month. The Christiansens had paid $75 per month for their apartment in Oak Park. Jack Christiansen would live on Bainbridge Island for the rest of his life.

EARLY CONCRETE SHELLS

Settled into Seattle, with his sense of adventure satiated at least for the moment, Christiansen began to refocus on his engineering career. He entered into the office at another time of transition for the firm. George Runciman had recently stepped back from management of the company, opening the door for another young engineer, named John Skilling (1921–98), to take the reins. Skilling was a dynamic entrepreneur—innovative, creative, and driven to expand the firm in new directions. He had spent the war working for Boeing, designing and implementing fixes to B-29 airplane designs. Afterward, he resumed his civil engineering education at the University of Washington, parking cars at night to pay the bills. Starting as an intern in August 1947, Skilling quickly established himself in the firm and began forming personal relationships with architects, owners, and contractors. He was equal parts engineer, salesman, and bartender—a potent combination of skills that would take the firm to unimaginable heights.[3]

As the old guard, Harold Worthington balanced Skilling's exuberance, with consistent office management and meticulous bookkeeping. He had grown up on a homestead in the wilds of the Olympic Peninsula, coming east to Seattle to attend the University of Washington in civil engineering. Hired directly out of school, Worthington spent his early years designing small, industrial projects, but he was thrust into a management position early on, following Witt's sudden death. As an administrator, Worthington navigated the firm through the Depression years—accepting the burden and responsibility to keep the lights on and the doors open. Though careful and reserved, with a Depression-era prudence, Worthington recognized that the 1950s were a time for growth. Beginning in 1947, he began rapidly hiring University of Washington graduates to help accomplish the work brought in by Maloney and NBBJ—increasing the size of the firm from just three employees in 1945 to fifteen in 1953. With Skilling continuing to bring in work and drive innovation and Worthington overseeing the internal workflow, the company was growing as fast as they could manage. Recognizing this balance, in 1955 they changed the name of the firm from the W. H. Witt Company to Worthington & Skilling.

Unlike Skilling, Christiansen was more measured and soft-spoken, but he fit into the office dynamic well. He preferred to focus on a building project itself, with less concern for the marketing or management side of professional engineering practice. With Skilling and Worthington keeping the firm going, Christiansen found himself with an incredible opportunity to just build and to continue his exploration of thin shell concrete structures.

Thin shell concrete was increasingly the subject of national publications. *Architectural Record* and *Architectural Forum*, as well as the regional *Pacific Architect and Builder*, began publishing articles on thin shell concrete projects in the early 1950s. Journals such as *Engineering News-Record* and *Civil Engineering* devoted more articles to the performance of shells and their transmission from Europe to the United States.[4] Seeing this publicity and learning about the intricacies of thin shell design, Christiansen began to recognize how the technique could be uniquely suited to building in the Pacific Northwest. Concrete was far cheaper than steel in the region due to high-quality cement manufacture in the North Cascades. With large stands of timber nearby, wood for formwork was readily available. Christiansen also found that although architects, engineers, and contractors were in separate firms (unlike in Chicago), collaboration among them happened easily in the Northwest—a sense of teamwork that was needed for thin shell concrete to work. Christiansen didn't know whether this was due to the region's distance from the large cities in the East (where entrenched professionals had already solidified specific roles), the influence of Boeing (which demanded a unity between design and engineering), or simply the frontier attitude of problem solving (using available resources to *get things done*). What he did know was that people were open to interdisciplinary collaboration and new ways of thinking about problems. Recognizing all these components, Christiansen began to investigate how thin shell concrete structures could be not only striking, expressive architectural forms but also structurally sound and economical to build.

ASCE MANUAL 31

A fortuitous event set Christiansen's thin shell career into motion. The same year he was hired at the W. H. Witt Company, the American Society of Civil Engineers (ASCE) published its first "manual of practice" addressing concrete shells, titled *Design of Cylindrical Concrete Shell Roofs*, known as ASCE Manual 31.[5] The ASCE, with the support of the Portland Cement Association, had organized a subcommittee on thin shell design in 1946, with the intent of simplifying the structural analysis of this type of shell, in hopes of encouraging its widespread use.[6] This manual, written primarily by engineer Alfred L. Parme, combined into a single design guide a wide body of shell analysis techniques that had been developed over the previous forty years. This included the shell theories developed by Dischinger and Finsterwalder for their Z-D system in the 1930s, as well as others developed since then.[7] But more than a theoretical treatise, this publication provided American structural engineers with a series of tables and charts that drastically reduced the time required to analyze and design a cylindrical concrete shell. This manual provided just the boost that Christiansen needed to start his exploration into thin shell concrete.

The manual presented the geometry of the cylindrical barrel vault in a simple, straightforward way, with only a limited number of variables. By definition, the cylindrical vaults have a constant radius (r) that defines the curvature of the shell, as well as a typically constant thickness (t). The arc of each constructed shell is thus a segment of a circle, defined by an angle in degrees (one quarter of a circle is 90 degrees). This arc is then extruded longitudinally, allowing the shell to span in a perpendicular direction (l). These simple variables helped Christiansen define his shell geometries in a precise way.

ϕ_k	h/ℓ	h/r	r/ℓ	s/ℓ	$\frac{I}{t^4}$
5°	.022	.0038	5.737	1.001	
10°	.044	.0152	2.879	1.006	
15°	.066	.0341	1.932	1.011	
20°	.088	.0603	1.462	1.021	.00018
25°	.111	.0937	1.183	1.032	.00068
30°	.1340	.1340	1.000	1.047	.00165
35°	.1577	.1808	.8118	1.066	.00355
40°	.1820	.2340	.7776	1.086	.00687
45°	.208	.2929	.707	1.111	.01216
50°	.233	.3572	.653	1.139	.02017

See Eshbach p. 1-28.

$$r = \frac{4h^2 + \ell^2}{8h} \qquad h = r(1 - \cos\phi_k)$$

1 radian = 57.296°

3.1416 $\pi^2 = 9.86965$ $\pi^3 =$ $\pi^4 = 97.4100$

Details from Christiansen's copy of the American Society of Civil Engineers' *Design of Cylindrical Concrete Shell Roofs* (ASCE Manual 31), published 1952

One important way of categorizing the different types of barrel-vaulted shells is through a comparison of the longitudinal span of the vault to the radius of its curvature (l/r). If the span is large compared to the radius, this ratio will go up, and the shell is determined to be a *long* shell with behavior more similar to a beam (carrying loads through bending between vertical supports). An approximate ratio for long shells is an l/r of 5.0. If the length is short with respect to the radius, it is called a *short* shell (l/r less than 1.0) and has behavior more similar to an arch (generating thrust between buttressed supports). Shells with a ratio in between are called *intermediate* shells, with behavior between that of a beam and an arch. Christiansen quickly discovered that a simple changing of the variables would produce a drastically different shell form.

Once the overall shell geometry was established, the design guide could help calculate the internal stresses within the shell and determine the size of the necessary edge beams or stiffeners. These edge beams—solid concrete sections running along the edges of the shell—helped stabilize the shell geometry and provided additional strength.[8] The manual separated the curved cylindrical shell from the edge beam within a design and allowed Christiansen to approach each individual project as a two-part process: the shaping of the shell geometry to match each project and the sizing of the edge beam to satisfy structural demands. With a strong background in structural analysis, he easily navigated the geometry and force calculations in the manual and used the tables to determine appropriate sizes of edge beams.

More importantly, however, Christiansen began to understand the spatial possibilities of the simple form. As he changed the span-to-radius ratio and the physical dimensions of a structure, he was able to create distinctly different types of architectural spaces. Soaring arches could create soaring overhead space, while long barrels could create more linear, horizontal spaces. Simply by adjusting the numbers, the cylindrical shell could become many different things. Christiansen's visualization of potential forms would be critical in his discussions with architects and essential to the future success of the barrel-vaulted shell.

While the manual provided the basis for design, it offered no guidance on the construction or forming of thin shells from concrete. To build the shells in the manual, Christiansen would have to figure out a formwork both strong enough to support the weight of the concrete until it could support itself and precise enough to give the shells their carefully defined geometry. Christiansen knew that building these shells would not be a simple task.

THE GREEN LAKE POOL

Christiansen's first opportunity to design a thin shell concrete structure came in March 1954—less than two years after he joined the firm—in the design of the Green Lake indoor pool facility. Like many civic structures, the indoor pool was long anticipated—advocated for by the citizens of north Seattle for over twenty-five years. In 1948, the City of Seattle approved a $2.5 million bond to fund the improvements of parks, overseen by the Parks Board. The city hired the local architect Daniel E. Lamont, of Lamont & Fey, to design the pool facility as an eastern addition to the existing Green Lake Field House. Built in 1928 (before the Great Depression), the Art Deco field house was an early reinforced concrete structure, with streamlined details and textured finishes. Lamont's addition needed to relate to the earlier structure, while expressing a new, modern, long-span space on a very tight budget. Known for its innovative work in reinforced concrete, Lamont & Fey hired the W. H. Witt Company for the structural engineering.

Within the W. H. Witt office, Christiansen saw an opportunity. He first approached Skilling and proposed using a thin shell concrete design for this project, and together they pitched the idea to Lamont & Fey. The primary design issue at hand was how to enclose and cover the swimming pool space in a low-cost manner, while relating the structure to the earlier field house. A structural steel option would have required large, deep beams, roughly 3–4 feet deep, or several large trusses. The only viable wood option would have been glulam beams, more than 4 feet deep, or heavy timber trusses. As Christiansen was discovering, thin shell concrete could be extremely economical and, with the right proportion, easily enclose the pool space with only a few *inches* of structural depth. Quite simply, the concrete shell required only a minimal volume of structural material, easily creating wide-open space and immense cost savings. If they could pull it off, the Green Lake Pool was a good candidate for a barrel-vaulted, reinforced-concrete shell.

The swimming pool was 75 feet long and 42 feet wide. To enclose the surrounding walkways and support spaces, Christiansen needed his shell to cover a space 110 feet long and 60 feet wide. With the length established (110 feet), Christiansen needed to decide what radius of shell to use and determine whether it would be primarily an arch or a beam. Using the ASCE design guide, Christiansen chose a shell with a 54-foot radius, swept through a 70-degree arc. With a 110-foot length, the shell had a span-to-radius ratio of roughly 2.0, making it an *intermediate* shell by the definitions in the design guide. With this configuration, the shell would need to be only 3.25 inches thick.

Four corner columns raised the entire shell 20 feet into the air to allow for entrances and windows. Concrete end walls closed off each end of the vault, and 38-inch edge beams spanned between columns in the long direction, infilled with concrete walls below. With this design, Christiansen had proposed a structural form of minimal, inexpensive material that fulfilled the simple, spatial requirements of the building's function. The rounded vault created a distinctive roofline that contrasted with the blocky field house form, clearly separate from the columns and enclosing walls. Though certainly approving of the building's form, Lamont (as the architect) played only a minor role in the external appearance—using 1-inch-by-8-inch boards to create a rough textured finish. As would be the case with many thin shell structures to come, Jack Christiansen was the one who played a primary role in determining the overall building form.

Completed Green Lake Pool, Seattle. Christiansen's first executed shell structure, 1954.

Even with the design in place, the question remained of how to build it. Christiansen knew that the formwork needed to shape the vault—a scaffolding both to support the weight of the wet concrete and to define the vault's geometry—was essential to the success of the project. From his review of other projects, he knew that if he wasn't careful, more than 50 percent of the entire project budget could be spent on formwork alone. With that in mind, Christiansen worked closely with the contractor, Cawdrey & Vemo, to set out a specific procedure to build the concrete vault in the easiest, least-expensive way possible.[9]

First, a full-size panel of the roof curvature was measured and drawn flat on the ground. Christiansen had included these measurable dimensions in the building plans. The arc was then framed out with dimension lumber—shaped to match the curve. This wooden arc was then used to form the top chord of a bowstring truss that would both delineate the geometry of the vault and support the concrete until it hardened. The contractor built several of these identical trusses, to connect together to form the entire barrel vault. The entire form was assembled vertically, like "something on the order of a toy building set."[10] The contractors then positioned a thin layer of reinforcing bars and delivered the concrete by large buckets, lifted by crane. Christiansen specified a particular concrete mix that would flow around the reinforcing bars, yet not slide down the slope of the form. The construction method was successful, and Cawdrey & Vemo completed the building on time and on budget.

In addition to filling a long-desired civic need, the construction of the Green Lake Pool attracted a great deal of local attention. The *Seattle Times* reported that the vault was an "unusual 'egg-shell' roof," the "first Seattle building of this design."[11] Other articles celebrated the pool as "a new type for this area."[12] Praising the innovation used in its design, *Pacific Architect and Builder* announced that the "cylindrical barrel thin shell concrete roofing has made its Seattle debut."[13] With this project, Christiansen had executed his first thin shell vault in the Puget Sound area. Skilling, the persistent promoter, already claimed that it was the largest, intermediate barrel vault in the world.[14]

SEATTLE SCHOOL DISTRICT WAREHOUSE

After the success of the Green Lake Pool, Skilling and Christiansen saw a huge potential for thin shell concrete. With the project coming in under budget, it became clear that thin shell concrete could indeed be a low-cost solution for enclosing large areas—but not every space could be covered with a single barrel-vaulted shell. If Christiansen and Skilling were going to expand its use, they would have to explore many different configurations and proportions—ones that matched different types of architectural spaces. They also knew that this complexity would require more detailed construction methods. Building the formwork for the Green Lake Pool had been a major component of the cost, only to have it disassembled and thrown away. If a form could be built and reused, then multiple shells could be cast at only minimal additional cost. Christiansen had learned from the Green Lake Pool experience and began to seek out other opportunities to use thin shell concrete.

Skilling and Christiansen approached a longtime architectural client, John W. Maloney (1896–1978), about using a thin shell structure in his next project: a warehouse for the Seattle School District in the Cascade neighborhood (the building was named the Central Supply Center). Maloney was one of the most prolific architects in the region at that time, with strong connections to the Catholic Church and other municipalities.[15] Earlier in his career, Maloney had designed in an eclectic style, but in the postwar period he shifted his attention to material expression and structure—a more modern approach that was an ideal match with Christiansen's creative structures.

The Seattle School District was expanding rapidly due to a large increase in student enrollment.[16] With an overall need for expanded facilities, the district specifically needed warehouse space to inventory its own equipment and supplies. It would be the first district warehouse constructed in over thirty years, during which time the student population had more than quadrupled.[17] For its warehouse, the district selected the site of the former Cascade School (Pontius Avenue and Harrison Street), which had been badly damaged in the 1949 earthquake and demolished (the playground was maintained as a city park). Despite the overall population growth, the Cascade neighborhood was transitioning from primarily residential to commercial building stock, as more families moved to the suburbs, and the district elected not to rebuild the school.

To store an inventory of books, cleaning supplies, lunchroom equipment, and desks, the district needed new warehouse space—a building ten times the size of the Green Lake Pool. In order to use modern freight-handling equipment, the building required large, clear spans and could have only a minimal number of columns. Christiansen knew then that thin shell concrete could provide an economical option.

Instead of the single vault he designed for the Green Lake Pool, Christiansen proposed a series of thin shell vaults with very different proportions from the shell he planned before. Each vault covered a space 33 feet wide (half the width as before) but extended long (128 feet in length). The shell had a constant radius of 25 feet, giving it a span-to-radius ratio of more than 5.0—making it a *long* shell according to ASCE Manual 31. That meant that these shells would behave structurally as long beams (rather than arches) and create a much more linear space. By arranging these shells both next to each other and end to end, Christiansen could create an expansive interior space, with the majority of columns around the perimeter and only one row of columns down the middle. End panels, split by long strip windows, filled in the end of each vault to provide stiffness in short direction. Working with Maloney, Christiansen came up with a scheme of eight concrete vaults (four vaults wide, two vaults long) to enclose the 49,000-square-foot main warehouse, attached to a light-steel-frame loading dock.[18]

The repetitive shell solution relied on a different approach to construction, with a method that enabled the formwork for one shell to be reused for another. Unlike the Green Lake Pool project, Christiansen designed a form-work system and sequence himself, including "suggested formwork details" on the construction documents for the contractor to reference. Given the legal separation of trades, Christiansen couldn't require the contractor to follow his suggestions, but he knew (as did others) that the success of the project relied on it.[19] Christiansen designed a series of rounded, lightweight wooden trusses that could be assembled in place and, through a coordinated sequence, used many times over. He described a process of jacking trusses into position and supporting them on timber shores and bracing. After the concrete was poured and set, the jacks could be repositioned, the shores removed, the trusses lowered, and the entire system rolled on wheels to the adjacent bay. Christiansen's construction suggestions were a part of the documents that contractors used to bid for the job.

Seattle School District Warehouse, Seattle. Under construction, showing Christiansen-specified movable formwork for segmental casting of barrel-vaulted shells, 1956.

Christiansen's attention to not only structural form and space but also construction details paid off grandly. The warehouse was sent out to bid to several area contractors, and the Seattle School District chose the Howard S. Wright Company. Upon reviewing the bid proposal, the district found the estimated cost to build the proposed warehouse so low that the district could afford to expand the number of constructed vaults from eight (as seen in the original drawing) to fourteen (the number finally built) and still remain under budget. The initial scheme, enclosing forty-nine thousand square feet, could be easily expanded to seventy-four thousand square feet. The Howard S. Wright Company planned to build only a single form and use it fourteen times over the course of several weeks.[20] The repetitive nature of Christiansen's design and a carefully coordinated construction process meant that the district could get more space for its money. When compared to the Green Lake Pool (acknowledging differences in mechanical, plumbing, and electrical demands), the project created ten times the enclosed space for only twice the cost.[21]

Seattle School District Warehouse, Seattle. Under construction, 1956. (*top*) Christiansen on top of cast vaults, inspecting concrete work. (*bottom*) Workers pouring concrete by bucket delivery.

Seattle School District Warehouse,
Seattle. Complete, 1956.
(*above*) View from street.
(*left*) Aerial view.

The structure was an industrial building, and Maloney, the architect, paid little attention to architectural finishes, but like Lamont before him, he utilized the unique spatial possibilities that the thin shell concrete offered. The repetitive structural vaults created a distinct roofline, which contrasted with the surrounding flat- and pitched-roof buildings. Maloney also detailed the vault edge to extend beyond the exterior columns and end walls to further express the thinness of the shell. Narrow strip windows running along the exterior controlled the natural light entering the building, while exhibiting the long-span capabilities of the shell. While the shell was likely chosen for its economic and structural performance, Maloney was clearly experimenting with the architectural possibilities that the structure allowed. In this modest, utilitarian structure, Maloney began to envision (as Christiansen

and Skilling had) an expanded use for thin shell concrete as a modern building form.

BUILDING CODE REVISIONS

Both the Green Lake Pool and the school district warehouse were built under
the 1943 Seattle Building Code.[22] Because the code had no provisions for
thin shell concrete, Christiansen and Skilling had to make their case for the
safety of these thin shell forms directly to the Seattle Building Department.

But broader changes were coming to the code. In 1953, Seattle mayor
Allan Pomeroy began a wide-reaching effort to update Seattle's building
codes—an effort his successor, Gordon S. Clinton, continued. The goal was to
replace the "outmoded" code with a "new, more flexible and better organized
law which would facilitate the growth" of Seattle.[23] More than two hundred
architects, engineers, contractors, and builders came together to rewrite the
majority of the code—something Clinton claimed was "unique in the history
of municipal code work."[24]

The resulting code was extremely advanced in its acceptance of new
structural engineering techniques. That was due, in part, to the chairman
of the committee in charge of all engineering—a structural engineer named
Peter H. Hostmark (1903–69).[25] Hostmark was a local, creative engineer
interested in designing with new techniques like prestressed and thin shell
concrete. In the decade to follow, Hostmark would design with both—most
prominently with architect Paul Thiry on the Washington State Coliseum
(now KeyArena, 1960–62) and St. Demetrios Church (1960–62).

As a result of Hostmark's involvement, the 1956 code contained provisions
for both prestressed concrete (sec. 2627) and thin shell concrete (sec. 2629).
While the prestressed concrete section was quite detailed, the thin shell concrete portion was limited to a single sentence. Simple, and broadly accepting,
it read: "Thin shell concrete construction (defined as structures in which loads
are sustained by thin, curved slabs) shall be permitted provided it is designed
and constructed according to established engineering practices."[26]

With minimal regulation of thin shell concrete, Christiansen had incredible freedom to design with the medium, so long as he justified his approach
through "established" practices. The code's language allowed Christiansen to
logically and responsibly design structures without fear of rejection by the
Seattle Code Committee—opening up even more possibilities of form and
space. The code was formally approved on September 11, 1956, and put into
effect thirty days later.[27]

Population growth was not limited to Seattle, and schools were expanding across the state. Maloney had a long history of designing schools in the Pacific Northwest, ranging from eastern Washington to Oregon, and won commissions for several new schools in the mid-1950s. Earlier in his career, Maloney's school designs were rather conservative, using a typical steel or concrete frame, with terra-cotta or brick detailing. But now, with Christiansen on board as an effective structural designer, Maloney began to change his architecture—motivated by the dramatic exterior profile and expansive interior spaces that thin shell concrete could offer. Typically, schools required two different types of spaces: repetitive, small-scale classrooms and individual, large-scale auditoriums or gymnasiums. Maloney began to see how these spaces could be designed with thin shell concrete, in an exciting, formally expressive way while still remaining economical to build. Christiansen's interest and capability had sparked a modern turn for one of the region's most prolific architects.

Their next project was to create an arts and music building, a gymnasium, and workshop buildings for a new high school in Ellensburg, Washington. Christiansen and Maloney dove headfirst into the opportunity to explore the barrel-vaulted form further and investigate the spanning solutions it offered. At this point, both Maloney and Christiansen knew that by significantly varying the radius of the shell with respect to its length, while also elongating different dimensions, the designers could create significantly different types of spaces. In terms of the design guide, the project would have a combination of *short*, *intermediate*, and *long* shells—utilizing their spatial characteristics to the fullest.

For the arts and music building, Maloney and Christiansen used a similar approach as the Seattle School District Warehouse. The building used multiple barrel vaults in sequence—covering an area 30 feet wide and 60 feet long, giving them proportions of an *intermediate* shell. This arrangement naturally divided the building into 30-foot bays, which accommodated individual classroom and practice spaces well. As with the Seattle School District Warehouse, Maloney detailed the shell to extend beyond the perimeter of the enclosure—making the 3-inch thickness of the shell a significant feature of the building's appearance.

To cover the long-span gymnasium, Maloney and Christiansen used a different configuration. They significantly increased the radius of the vault to 141 feet to span 175 feet in width, while effectively shortening its length to only 19 feet, making the structure a *short* shell. This length dimension

Ellensburg Senior High School, Ellensburg, WA. Arts and music building vaults under construction, 1955.

measured the distance between thicker, stiffening ribs (16 inches wide by 24 inches deep), which arched the width of the vault. These ribs acted similarly to the edge beams of the longitudinal vault—collecting the loads from the shell and channeling them to the foundation elements and ground. The shell segment between the ribs stopped short of the ground—allowing for doorways, windows, and a vertical enclosure plane at the base. Five of these vaulted segments (each composed of a shell between ribs) together made up the entire gymnasium—a total length of almost 100 feet. With the top of the vault 36 feet off the ground, the gymnasium could accommodate sporting events and large assemblies of people. Maloney and Christiansen had created a distinctly different kind of space through the simple variations of the barrel vault geometry.

The workshop space was of nearly identical design, constructed right next to the gymnasium and connected through a shared foundation. Slightly smaller in scale, the shop space had a radius of 84 feet, with a reduced span of 104 feet, and rose 23 feet off the ground.

For the first two thin shell projects, formwork had been cut and assembled on-site, overseen by the general contractor. For the larger, more complex shells (e.g., the Ellensburg gymnasium and workshop), the formwork required an even greater level of expertise. For this work, Christiansen brought on a formwork-specific contractor (Timber Structures Inc. from Portland, Oregon) to build and operate movable, reusable forms that could speed up the casting of the shells.

Ellensburg Senior High School, Ellensburg, WA. Gymnasium and workshop arches under construction (arts and music building complete beyond), 1955.

Again, the economy of the entire school project was based on the repetitive use of the forms. The gymnasium was poured in 19-foot sections (between ribs)—one strip at a time, which meant that only one segment of formwork had to be built. Timber bowstring trusses supported wood joists sheathed in plywood to form the underside of the thin shell. When the concrete had set (after one or two days), the timber trusses were lowered 12 inches, rolled on I-beam tracks to the adjacent strip, and lifted into place. The same process was used for the attached workshop building. The arts and music building was cast similarly to the Seattle School District Warehouse.

TENSIONING STEEL IN YAKIMA

The formwork made for the Ellensburg Senior High School project led to a further extension of thin shell concrete buildings. The Ellensburg forms (from both types of barrel-vaulted spaces) were built not only for multiple uses on that job but with the intention to use them again on a completely separate project. Maloney and Christiansen were simultaneously designing buildings for the West Valley High School in Yakima, Washington, and, by hiring the same formwork builder, were able to reuse the formwork in the creation of another gymnasium and workshop building. This repetitive use of forms made it easy to produce inexpensive buildings that were functional, durable, and, above all, Modern.

While efficient in the thickness of the shell, these early barrel vaults still required deep edge beams or stiffening ribs along the shell's borders. These thick elements stiffened the edges of the shell, reducing the tendency to buckle under heavy loads. These edge beams and ribs served as boundary elements, required to collect all the loads from the shell and channel them to the foundation.

With a continual eye toward efficiency, Christiansen began to explore ways to reduce the size of these elements. To this end, he began to incorporate another emerging structural engineering technique, prestressing: the pre-tensioning and post-tensioning of steel cables within reinforced concrete. The two terms refer respectively to embedding steel cables within the concrete (anchored at each end) and applying an external tensioning force—either before the concrete is cast or after. This tensioning forces the concrete into compression, minimizing the net tension and splitting failure that causes concrete to crack. Drawing from the stellar engineering community that surrounded him, Christiansen took a night class at the University of Washington from prestressing pioneer Arthur R. Anderson (1910–95) to learn how to design with tensioned cables.[28] Christiansen learned how the additional compression provided by the cables meant that the edge beams of his shells could be smaller, and his concrete structures could be even thinner and lighter than ever before. Wherever high-tension forces were expected in concrete, Christiansen could use pre-tensioning (tensioned cables before the concrete was cast) or post-tensioning (tensioned after the concrete was poured) to counteract it. Christiansen recalled the introduction of prestress into his designs: "Prestress literally liberated concrete as a construction material, got rid of the cracks. The problem with concrete is it always cracked, but all of a sudden you got a material that is so durable that it will last thousands of years and it won't crack."[29]

In 1956, Christiansen brought together his growing expertise in thin shell concrete structures with prestressing techniques for the first time in a gymnasium for the Wilson Junior High School in Yakima (again with Maloney). For multiple cylindrical vaults—41 feet wide and 135 feet long, Christiansen used post-tensioning in the edge beam at the base of the vault, significantly reducing the size of the dividing beams. The smaller beams lowered the self-weight of the beams and allowed even more natural light into the interior spaces.

In 1957, Skilling and Christiansen publicized their work at a regional conference for the American Concrete Institute (ACI) (held in Seattle, November 4–6), in a paper titled "Prestressing of Cylindrical Concrete Shell Structures."[30] In their presentation, they described the advantages of prestressing cylindrical concrete shells (to reduce edge beams' size) and the methods they used. While in Europe and England other designers, including Finsterwalder, were utilizing prestressing, Christiansen believed this shell to be the first in the United States to have a cast-in-place, post-tensioned edge

Wilson Junior High School, Yakima, WA. Gymnasium with post-tensioned concrete edge beams. (*top*) Under construction, 1956. (*bottom*) Complete, 1956.

beam.[31] At that time, the prestressing pioneer T. Y. Lin was also interested in this technique, using prestressing in the construction of a folded-plate shell for La Rinconada Hippodrome (1959) in Caracas, Venezuela.[32] And it was not until two years later that Lin published his article titled "New Design Horizons through Prestressing of Concrete Shells."[33]

By combining the outward, form-based strength of thin shells with the internal (nearly hidden) strength attained from prestressing, Christiansen's structural designs became thinner and thinner, while achieving larger and larger spans. Not content to simply repeat the same structural design, he continually sought to innovate and improve his work. He balanced careful calculation of structural forces with an architectural appreciation of the changing thin shell forms. Harold Worthington and John Skilling continued to support Christiansen and encouraged him to experiment and innovate. Speaking collectively, Skilling recalled: "We had to be dreamers to do thin shells in the early 1950s. There was no one to tell us what to do, so we had to work out answers for ourselves. This pioneering didn't make us any money, but it taught us what we needed to know about thin shells. We developed our concepts intuitively, analyzed them, and designed those structures conservatively."[34] The use of these thin shell concrete vaults throughout Washington State was prolific and began to attract national attention. In January 1956, *Engineering News-Record* reported that "the design appears to have caught on more quickly in Washington than in other parts of the country" and that "the Yakima area probably will have more schools of thin-shell construction than any other area in the country."[35] The economics of thin shell concrete construction paired with an expressive architectural form and adaptive space capabilities had proved to be a powerful combination.

CLIMBING ADVENTURES

During this period of exploration with thin shell concrete, Christiansen was also living out his dream of climbing the mountains of the American West. In 1953, he (and Sue) enrolled in a mountaineering course offered by Olympic Community College in Bremerton, where they learned about the surrounding mountain ranges and the basic techniques of climbing. Christiansen first started climbing peaks in the Olympic Mountains, as they were close enough to his home on Bainbridge Island that they could be climbed on a weekend trip. In 1956, Christiansen summited Mount St. Helens for the first time—a peak with particular significance in the Northwest. Christiansen fell in love with the entire process of mountaineering—the ambition to summit an unknown peak, the planning, the technical understanding of the gear required, the uncertainty, and the perseverance of reaching the destination.

Christiansen's growing passion for climbing mirrored his expanding work in thin shell concrete. Just as his increasing ability as a mountaineer took him to taller and taller peaks, Christiansen's growing expertise in shell designs was leading him to more and more ambitious projects. While his earlier school projects used mid-scale repetitive vaults, Christiansen knew that the barrel-vaulted form could be used for more individual, signature buildings as well. With careful designing, he could push the span of a thin shell further than ever before and yet also create more sensitive architectural space.

Jack Christiansen (*far right*), with climbing partners on Mount St. Helens, 1956

INCREDIBLE SCALE: THE BOEING B-52 HANGAR, MOSES LAKE

With a solid understanding of their mechanics and behavior, Christiansen began confidently extending the scale of his structural designs, opening up the opportunity to design a new type of building: hangars for airplanes. The Boeing Aircraft Company had expanded its facilities to Moses Lake, Washington, establishing an airfield for further testing of the B-52 Stratofortress—a long-range bomber with a wingspan of 185 feet. Boeing had just built a steel-truss hangar for three B-52 airplanes in 1955 but, with expanding orders, needed an additional facility to house eight more B-52s, along with support shops in between.

The architectural firm selected for the project was NBBJ. This firm, like Maloney's, was a longtime client of the W. H. Witt Company, but had not experimented with thin shell concrete before. It had designed several airplane hangars for the military and for Boeing but had typically used steel trusses to span the long distances—often designed by other engineers at the W. H. Witt Company. The lead architect on the project, William Bain Sr., was intrigued by Christiansen's growing expertise, but had reservations. The challenge was to find a structural system that could allow planes to enter the hangar from each side and park nose to nose. The size and profile of the planes required an internal clear span of at least 215 feet (for the wingspan) and an overhead clearance of 65 feet (to accommodate the tail). Thin shell concrete was great for low-rise, repetitive structures, but could it be designed at this massive scale? How could a cylindrical vault accommodate the geometry of the plane?

To meet these demands, Christiansen developed a hybrid structure composed of a series of thin shell cylindrical vaults, supported by post-tensioned concrete piers. These piers elevate and separate the thin shell vaults, creating an undulating profile—"like ocean swells."[36] The clear span profile of this structural system efficiently matched the profile of the B-52 airplane, reducing wasted space overhead. Four large cylindrical vaults, each arching 135 feet, were stiffened by a series of six deeper arches, 66 feet apart. These arches were then pin connected to the ends of cantilevered, post-tensioned supports that brought the vault loads to the ground. The arch and vault thrusts were balanced through the interior of the structure with buttressing walls and taken by stiff frames at each end. This combination of shells produced an overall clear span of 220 feet.

B-52 airplane hangar for the Boeing Aircraft Company, Moses Lake, WA. Single bay under construction, 1956.

Christiansen used his experience on the smaller Seattle School District Warehouse to design an efficient, reusable formwork system. The contractor was the Howard S. Wright Company—the same firm that had built the district warehouse. Each shell form supported a roof segment 66 feet long and was used a minimum of four times. Once again, Christiansen provided detailed construction notes on the contract documents, describing a specific procedure for casting, lowering, and moving the forms—this time on steel tracks to different locations.[37]

Still uncertain about the suitability of thin shell concrete, Boeing asked Christiansen to prepare a complete alternate design using steel trusses instead. The company needed assurance that this thin shell design wasn't going to be too expensive and wanted a backup plan. Christiansen and Skilling quickly drew up a steel-truss design of a profile similar to the thin shell, but nowhere near as elegant. The Howard S. Wright Company evaluated the costs of the two designs and determined the difference between them to be only $14,000, of a $5.8 million project. Even though the thin shell concrete solution was *slightly* more expensive, it was chosen over steel for two other reasons: a better resistance to fire and less maintenance required. Thin shell concrete became the preferred option for its performance and long-term durability.

B-52 airplane hangar for the Boeing Aircraft Company, Moses Lake, WA. Complete, with Christiansen out front, 1956.

Once completed, the hangar attracted significant national attention. *Engineering News-Record* wrote that the hangar's roof spanned "big bays."[38] With an overall clear span of over 220 feet, the building was on the scale of some of the largest concrete structures in the world.[39]

EXPRESSIVE ARCHITECTURE: ST. EDWARD CHURCH

In addition to long-span applications, Christiansen demonstrated a careful ability to integrate thin shell concrete barrel vaults into more sensitive works of architecture. In 1956, Maloney was commissioned to design a new Catholic church in the Hillman City neighborhood, south of downtown Seattle. Even though Maloney had designed with thin shell concrete for warehouses and schools, he now sought to use a thin shell concrete vault for a sacred, religious space.

Inspired by the thin lines of the shell as a medium of Modern architecture, Maloney designed the primary sanctuary space to be enclosed by a single, cylindrical shell. As he had done with the Seattle School District Warehouse and the Ellensburg Senior High School, he extended the vault beyond the end walls and over the longitudinal walls on each side. With this treatment, the shell seemed to wrap the building, embracing the space within, becoming the dominant feature of the church's design.

This simple, sensitive design required a unique type of shell. Measuring 43 feet wide (with a radius of 37 feet) and 153 feet in length, the shell was one of the longest-spanning vaults in the United States. In addition, Maloney did not want the typical solid edge beams that this shell required, seeking instead greater transparency and light through windows around the base of the shell. To accommodate this plan, Christiansen designed large, rectangular openings within the depth of the beams—essentially turning them into Vierendeel trusses.[40] Though technically over 11 feet deep, the edge beams were so full of openings—and stained glass windows—that they dissolved as filtered light filled the sanctuary space along its entire length. These edge members still carried the loads from the shell as well as the lower roofs over the aisles, allowing clear-span access to the sanctuary space from the connecting buildings. The end walls (typically solid concrete in earlier projects) were accentuated with precast stone mosaics, depicting the patron saint, St. Edward.

The shell for St. Edward Church attracted national attention from the ACI for its long span and unique edge beam design.[41] *Pacific Architect and Builder* claimed that it was the largest single thin shell concrete design of the time.[42] More importantly, however, the shell was the defining spatial element of a carefully designed community church—an architectural/structural component of a quiet, reflective space. In this project, the thin shell concrete vault was used not only for its economy and structural performance but also for its space-shaping ability.

Between 1956 and 1958, Christiansen used cylindrical shells—some prestressed—in over twenty-three structures around the state, all slightly varied in their size and configuration. Educational buildings included West Valley High School, Yakima (1956, gymnasium and classroom); Asa Mercer Middle School, Seattle (1957, gymnasium and classrooms); Highland Junior High School, Bellevue (1957, classrooms and gymnasium); Quincy High School,

St. Edward Church, Seattle. Completed 1958, contemporary view.

Quincy (1957, gymnasium); and Wapato High School, Wapato (1957, gymnasium). Chief Sealth High School in Seattle (1958, gymnasium, classrooms, and theater), designed with NBBJ, attracted particular attention for Christiansen's design of a single-shell enclosing a 110-foot-by-150-foot auditorium and for the efficient use of formwork for the classroom spaces.[43] Industrial structures included a Washington State Penitentiary shop building in Walla Walla, with the barrel-vaulted shells opened to accept northern light. And Jack Christiansen was just getting started.

CHAPTER 3

GEOMETRIC COMPLEXITY

Design with Hyperbolic Paraboloid Shells, 1958–1966

T HE SPREAD OF barrel-vaulted, cylindrical shells across Washington State was remarkable and proof that thin shell concrete was an economical yet innovative structural system. Creatively restless, Jack Christiansen soon began exploring other shell geometries as well. Inspired by the published work of Spanish-Mexican engineer Félix Candela, Christiansen adopted the hyperbolic paraboloid—a warped structural surface. With curvature in two directions at once, the hyperbolic paraboloid offered not only a different structural behavior but also new spatial opportunities for architecture. Christiansen soon convinced his architectural clients to experiment in this innovative form and found even further success.

IN THE OFFICE

During this time, Christiansen had also established himself as a creative and productive engineer in the Worthington & Skilling office. This working environment provided both the organizational framework and the professional opportunities for him to thrive. Worthington & Skilling had its offices in the Lloyd Building, at the corner of Sixth Avenue and Stewart Street in downtown Seattle. In the mid-1950s, the firm had a small number of employees (roughly fifteen) but was beginning to hire more to meet an expanding building market.

Harold Worthington, as managing partner, maintained a tight office culture with Depression-era thriftiness. The majority of the engineers in the office had been educated in the Department of Civil and Environmental Engineering at the University of Washington. The engineers (all men at the time) were expected to wear white shirts, black ties, and dress pants every day to the office. Each engineer was assigned a seat between two desks—an angled drafting table on one side and a flat calculation table on the other—and expected to stay at his desk all day. Only scheduled breaks for coffee and lunch were permitted. Insisting that all engineers use their drafting pencils down to the end, Worthington provided them with "pencil extenders"—metal attachments that clipped to remaining pencil stubs (too small to hold)—and helped the engineers use every last bit of pencil lead. On one occasion, Worthington had quickly lifted a sheet of drafting film off Christiansen's desk, sending his pencil flying out an open window. Christiansen laughed as his pencil fell to the street below, until he realized that Worthington was motioning for him to retrieve it.

Christiansen appreciated the rigid structure of the office. He found the environment effective in keeping him focused on creating detailed, efficient designs: if he wasn't wasting pencil lead at his desk, he certainly wasn't going to waste concrete and steel in his buildings.

At the same time, other engineers in the office were always willing to assist Christiansen in the execution of his creative work—each with his own area of expertise and a wealth of design experience. Joe Jackson, one of the older engineers, had worked extensively on industrial buildings. Worthington had specialized in concrete design in the 1930s. The office of Worthington & Skilling was essential to Christiansen's success in thin shell concrete.

With the economy still booming in the late 1950s, projects continued to flow to Christiansen's desk, presenting more and more opportunity to design. Creatively, however, Christiansen became interested in branching out beyond the cylindrical barrel vault. He had created a variety of buildings and spaces simply by adapting the forms contained in ASCE Manual 31, but he knew that thin shell concrete could be used in many other geometries and configurations. John Skilling, as a young, entrepreneurial leader, encouraged Christiansen to explore new designs—always interested in presenting something new to his architectural clients. Christiansen knew, however, that whatever he tried next must have a unified logic of construction and structural form. Free-form shells would be far too costly to build in the Pacific Northwest, so more than anything, Christiansen needed a new system to design within.

Inspiration soon came in the mail. The office subscribed to several professional publications, including the journals of the ACI and the ASCE. These publications helped connect Christiansen and the other engineers in the office to industry changes and national trends, but they were specifically crucial in connecting Christiansen to the evolving national conversation around thin shell concrete design. The early 1950s were a particularly important moment in the growth of thin shell concrete in the United States, and journals published many different articles on thin shell structures.[1]

In January 1955, Christiansen read an article by Félix Candela on a different type of thin shell structure—one that Christiansen had never seen before.[2] Candela had earlier presented his work to the ACI's fiftieth annual convention in Denver, in February 1954, and a similar paper at a conference on thin shell concrete at MIT, in June 1954.[3] In both venues, Candela distinguished himself from other engineers, architects, and builders by his ability to move seamlessly between topics of aesthetics and space, to means of construction and calculation of structural forces. Through the article alone, Christiansen could see right away that Candela was doing something remarkable with his thin shells of concrete.

HYPERBOLIC PARABOLOIDS

Candela was championing a different geometrical form for thin shell structures—a hyperbolic paraboloid. The hyperbolic paraboloid defines a doubly curved surface, one that has curvature in two directions at the same time—a "saddle"-shaped surface. The surface could be equally defined two ways: either as the intersection of two perpendicular parabolic curves (curving away from each other) or as the twisting (or warping) of an otherwise flat surface. With this geometry, Candela's thin shell structures, like the Cosmic Rays Pavilion (1951) in Mexico City, swooped and curved in three dimensions, creating striking, exciting buildings with many different types of architectural space.

But Candela's interest in the hyperbolic paraboloid extended beyond the forms and spaces it created—to how it was built and how it carried structural loads. Candela recognized that the rather complex geometry of the hyperbolic paraboloid could be formed from a relatively simple process of rotating, or twisting, a series of straight-line formwork—effectively "warping" a flat geometry into a twisted one. That meant that these shapes could be easily formed from traditional construction materials—such as dimension lumber or linear pieces of steel or aluminum. The concrete could be poured on top of the twisted formwork and take on the doubly curved shape as a thin shell. This simplicity meant that builders no longer needed to build

Félix Candela's Cosmic
Rays Pavilion, Mexico
City. Completed 1951,
contemporary view.

rounded formwork—as was typically required for all cylindrical vaults. Christiansen could see how this geometry could potentially reduce the labor and material costs associated with thin shell construction.

Candela was also an engineer and concerned with how these cast thin shell forms would work as load-carrying structures. Candela presented a simplified calculation method to calculate the forces within hyperbolic paraboloid shells. This analysis treated the shells as thin membranes and used simple calculations that took advantage of the geometry of the shell. While calculating the forces within a warped thin shell surface was complicated, doing it for a single curve (or parabola) was straightforward. By understanding that hyperbolic paraboloid surfaces were composed of just two intersecting curves, Candela showed how engineers could do two simple calculations (of curves) and combine them to approximate the forces within the thin shell surface. These calculations revealed that the hyperbolic paraboloid (when properly arranged) was also extremely efficient in carrying structural loads— with only 1–2 inches of concrete needed to span long spaces.

Christiansen was impressed by the range and diversity of the thin shell structures that Candela discussed in his article. By the time it was published, Candela had completed over fifteen hyperbolic paraboloid projects.[4] Christiansen saw that Candela had addressed many of the important aspects of thin shell construction that he himself had recognized on his own design of cylindrical vaults—structural analysis, formwork, construction procedure, and overall building form. This new geometry—the hyperbolic paraboloid—could potentially satisfy all the construction and engineering-related concerns, at a very low cost, while bringing new formal possibilities to buildings. Truly inspired, Christiansen began experimenting with the form right away.

WENATCHEE JUNIOR HIGH SCHOOL

Christiansen first experimented with hyperbolic paraboloid shells on the design of the Wenatchee Junior High School in Wenatchee, Washington, with the architects at Naramore, Bain, Brady & Johanson (NBBJ).[5] The design was completed, and construction documents signed, in October 1955, just ten months after Christiansen read the Candela article. On this project, Christiansen and the architects experimented with the spatial qualities of different thin shell concrete geometries, using both the cylindrical shell geometry and the new hyperbolic paraboloid form. This experimentation is most evident in three significant parts of the school project: the classroom buildings, the walkway coverings, and the gymnasium space. These elements show a progression away from the tried-and-true, single-curvature cylindrical form and into the world of double curvature, with new spatial possibilities.

To cover the classroom buildings, Christiansen and the architects at NBBJ used the familiar repetition of cylindrical shells—a single-curvature vault covering a single classroom space—but introduced a variation that created a different overall appearance. Each 3-inch-thick shell spanned just over 28 feet (as a barrel vault) over columns that defined the width of each classroom. Yet unlike for other schools Christiansen had designed, these classrooms were separated by 10-foot-wide walkways, covered with a shell of opposite curvature (curving upward). This arrangement gave the overall composition the look of an undulating wave in the landscape. Developed in collaboration with the architects at NBBJ, the building still retained the repetitive spaces that enabled a systematic, cost-effective forming system, while indicating an expanding design potential in thin shell concrete and a new design thinking.

This inventiveness is even more evident in other buildings on the school's campus. For the walkway covers between classrooms, Christiansen designed freestanding, hyperbolic paraboloid umbrellas—ones similar to Candela's in Mexico. Four, warped hyperbolic paraboloid panels surrounded a central column, providing an umbrella-like canopy over the walkways. Christiansen

Wenatchee Junior High School (now Pioneer Middle School), Wenatchee, WA. Completed, 1957. (*above*) Classrooms showing undulating cylindrical concrete vaults, contemporary view. (*left*) Christiansen's first use of the hyperbolic paraboloid, completed.

described these as "shell panels" (not shell vaults), with only modest dimensions compared to Candela's work in Mexico. Each four-panel assembly was 19 feet wide by 19 feet long in plan (each panel was 8.5 feet by 8.5 feet), with the center point dropped 2.5 feet to connect to a central circular column. As a result of the modest size, the shell was only 1.5 inches thick, increasing to 4 inches at the edges. Christiansen designed roughly fifteen of these open-air umbrellas around the classroom and administration buildings. These walkway covers were the first documented use of the hyperbolic paraboloid in thin shell concrete in Washington State.

In these designs, it is also evident that Christiansen did not just mimic Candela but began to design with the form in his own way. Surrounding the gymnasium, he used the hyperbolic paraboloid in a different way—as half shells (two-panel assemblies) that were supported by the walls surrounding the perimeter of the gymnasium. These panels essentially cantilevered from the wall, providing an overhead canopy along the edge of the gymnasium wall. Though simple, the use of shells in this way indicated how Christiansen would soon take the form in his own distinct direction.

The gymnasium is truly transitional—a design between the singly curved barrel vault and the doubly curved hyperbolic paraboloid. Christiansen and the architects at NBBJ appear to have begun with a single-curvature cylindrical shell but modified it to include double curvature. The structure spanned 174 feet across the short direction of the gym with a thin shell of concrete and regularly spaced stiffening ribs, yet in between the stiffening ribs Christiansen introduced a curvature in the perpendicular direction. This shell in between the ribs curved upward—creating only a slight rise (just about a foot) above the top surface of the ribs. This curvature created a pillowed-like appearance to the top of the shell—a bowing outward of the shell between the ribs—and gave the shell a double curvature.

Wenatchee Junior High School (now Pioneer Middle School), Wenatchee, WA. Gymnasium under construction, 1957.

Wenatchee Junior High School (now Pioneer Middle School), Wenatchee, WA. Gymnasium complete, 1957.

This doubly curved geometry, however, was not a hyperbolic paraboloid. To create this additional curvature, Christiansen needed a second set of rounded, trussed forms, placed in between the long-span trusses, and did not take advantage of the straight-line forming. This curvature was an experimental move (Christiansen did not repeat this geometry again), but it further illustrates his investigative process with new shell forms.

Not recognized or written about at the time, this gymnasium was Christiansen's first, large-scale structure that used a doubly curved concrete shell. As an original, creative use of thin shell concrete, Christiansen distinguished his work from Candela's while embarking on a newfound freedom of thin shell form.

EXPERIMENTATION IN HYPERBOLIC PARABOLOIDS

This experimentation in Wenatchee, however, did not spark more use of the hyperbolic paraboloid form. For schools that Christiansen designed with NBBJ later in 1956, he returned to a strictly barrel-vaulted arrangement, with repetitive cylindrical vaults over the classrooms and individual, long-span vaults over the gymnasiums. For Highland Junior High (designed by

NBBJ and Christiansen) in 1957, the design used no hyperbolic paraboloid forms at all—only cylindrical vaults (of varying sizes). For Chief Sealth High School (also designed by NBBJ and Christiansen) in 1957, the team used only cylindrical vaults. It appears that despite their mutual exploration of double curvature, the architects at NBBJ were not yet committed to using hyperbolic paraboloids on their other projects.

Christiansen still had some persuading to do. He needed to find the right way to convince his architectural clients that the hyperbolic paraboloid could be a rational yet exciting building form. In the office, he experimented with different arrangements of shells and worked on devising an efficient, repetitive construction method to keep costs low. It is possible that the form seemed too flamboyant or extreme to architects, at first glance. Indeed, the hyperbolic paraboloid did not have the historical connection that the barrel vault did, and that may have made the geometry seem more obscure or abstract. During this time, Christiansen worked hard to show different architects how the hyperbolic paraboloid could shape space in a unique way, but he needed an architect willing to truly explore a new type of form—not just the modification of an existing one.

Christiansen soon found a pair of young architects who were immediately interested in the expressive, thin shell forms, at the firm of Bassetti & Morse. Fred Bassetti (1917–2013), a Seattle native, graduated from the University of Washington with a bachelor of architecture in 1942 and then went to Harvard for a master of architecture (received in 1946).[6] He returned to Seattle and worked with NBBJ for roughly one year before forming his own firm in 1947 with Jack Morse.[7] Morse (1911–2000), who was a native of New England and held a master of architecture also from Harvard, came to Seattle in 1942. He started a partnership with the architect Ralph Burkhard in 1945, before joining with Bassetti.[8]

With several award-winning residential projects in the early 1950s, the architects at Bassetti & Morse had become a part of a new generation practicing in the Pacific Northwest.[9] They were young designers, trained in the early modernist tradition and now defining a regional modern architecture for the Pacific Northwest.[10] Bassetti and Morse were not transitioning from a previous architectural approach (as Maloney and the older partners of NBBJ were) but were exploring modernist ideas learned in school and open to new design directions. Like other designers in the region, Bassetti and Morse embraced expressive structural forms in their architecture, in both residential and school projects. Morse has even been described as a "structural rationalist" for his use of structural form.[11] This interest in structure is evident in many of their small-scale residential projects, through the exposed wood beams in interior spaces and the use of structural framing to define exterior details.[12] As the scale of their projects increased, they began to use different structural materials—like concrete—and again looked to these materials for architectural inspiration.

MERCER ISLAND HIGH SCHOOL MULTIPURPOSE ROOM

In fall 1957, Bassetti & Morse was commissioned to design a large multipurpose building for Mercer Island High School. The building needed to accommodate a large open assembly/cafeteria space and several smaller associated spaces, including a kitchen, a teachers' break room, a ticket booth, a platform stage, and storage. All the smaller spaces needed to surround and serve the larger, open space.[13]

To cover all of these functions under one roof, the architects needed a structure that could span over 140 feet. The smaller-scale projects that Bassetti & Morse had designed before could be easily spanned with conventional materials, but this space would need a different solution. This larger scale also meant that the overall structural form would play an even larger role in the architecture than in previous projects. Looking for a creative structural engineer to collaborate with, Bassetti & Morse turned to Jack Christiansen.[14]

Bassetti & Morse presented its architectural ideas for the multipurpose room, as separate functions under one roof. Christiansen saw a synergy between the architectural concept and the hyperbolic paraboloid forms he had been experimenting with in the office. Just as the program elements could come together to create an interior space, different panels of hyperbolic paraboloid shells could come together to create a singular building form. With a structural strategy that aligned with the programmatic demands, different shell segments could contain different program elements.

Together, the design team developed a centralized, 142-foot-diameter, dome-like space formed by seven identical hyperbolic paraboloid panels.[15] Vertical, upstanding ribs traced the boundaries between these panels, stiffening the structure while visually articulating its ruling geometry. These seven segments would help separate the different auxiliary functions, while still surrounding a central open space. The architects and Christiansen intentionally chose an odd number of segments, as it seemed to be "more dynamic" than an even number, avoiding a more classical biaxial symmetry composition.[16] Though governed by the spatial definition of the hyperbolic paraboloid, the architects and Christiansen also shaped the shell geometry to accommodate entrances, internal clearance heights, external form, and

Mercer Island High School Multipurpose Room, Mercer Island, WA. Formwork showing ruling geometry for concrete shell, 1958.

drainage. To accomplish that, Christiansen actually combined two surfaces of double curvature to create each of the seven panels. He explained: "Each of the seven segments is composed of two half-saddles of different curvature joined to produce a single, unsymmetrical shell. This increased [over]head height under the shell near the abutments."[17]

Christiansen combined two distinct shell geometries into one, along a shared line of parabolic curvature—creating a particularly complex geometry.[18] Christiansen worked with architects Bassetti & Morse to contour the edges of the shell, trimmed to accentuate the peak. Through this collaboration, the building became not just abstract geometry, but a shaped form—attuned to its architectural aesthetic. This manipulation of shell form was celebrated in local publications. *Pacific Architect and Builder*, in its December 1959 issue, wrote: "Variety of form and contrast of heights, together with the changing daylight brought in through the main entrances and over the balcony, create a lively, cheery building."[19]

Mercer Island High School Multipurpose Room, Mercer Island, WA.
(*left*) Casting concrete shell and stiffening ribs, 1958.
(*below*) Complete, 1958.

As a ruled surface, the shell was easily constructed with only straight beams, joists, and planks. As he had done before—in what was quickly becoming a trademark of his work—Christiansen provided explicit, clear instructions to the builders on how to form the complex geometry.

In this project, Christiansen showed an extremely complex and inventive use of the hyperbolic paraboloid. The Mercer Island Multipurpose Room was an exercise in geometric ingenuity that, in creating a unique structural-spatial condition, was unlike anything seen before in the Pacific Northwest and rivaled the work of Candela in Mexico. Widely written about, this structure reignited interest in hyperbolic paraboloids among local architects.

INGRAHAM HIGH SCHOOL

The Mercer Island Multipurpose Room was a breakthrough building for Christiansen. Through Bassetti & Morse's design, the building showed that hyperbolic paraboloids could create interesting, dynamic architectural space, yet maintain a low cost of construction. It indicated that hyperbolic paraboloids could play a role in the regional modern architecture of the Pacific Northwest.

Following this building's construction, the architects at NBBJ came to see the potential of the hyperbolic paraboloid for long-span, signature spaces. They began to see that the new formal language was not a variation of the barrel-vaulted form, but something completely different. In 1958, six months after the Mercer Island Multipurpose Room was completed, Christiansen teamed up again with NBBJ in the design of another high school—Ingraham High School in north Seattle. Christiansen designed cylindrical barrel vaults over a gymnasium space, with a unique prestressing arrangement. For these shells spanning 115 feet, Christiansen placed draped prestressing strands within the 3-inch thickness of the shell—not just in the lower edge beam. This technique was cutting-edge for the time, preceding August Komendant's prestressing of the Kimbell Art Museum vaults by ten years, and matched only by activities in Europe.[20]

Christiansen used a true hyperbolic paraboloid structure to cover the auditorium space. The overall form was composed of three hyperbolic paraboloid panels, again stiffened by upstanding ribs where they intersect, and all together spanning 160 feet between three supports. The 3-foot-thick shell rose to 28 feet at the apex. In March 1958, *Pacific Architect and Builder* published an image of the model, describing the building as "a combination of three hyperbolic paraboloids, with only three points of support."[21]

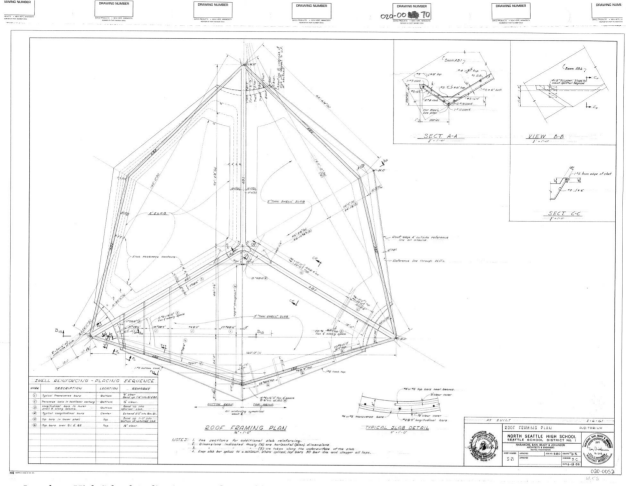

Ingraham High School Auditorium, Seattle. Roof framing plan, drawings dated 1958.

Later, in June 1958, the *Seattle Times* published a sketch showing the layout of the school and an accompanying description.[22] In describing the barrel-vaulted gymnasium and the hyperbolic paraboloid auditorium, architect Perry Johanson, from NBBJ, explained: "Both roof types are the cheapest known to cover the extensive areas involved. The roof is a shell of thin concrete with low maintenance cost."[23]

Yet after the building's completion in 1959, the *Seattle Times* called the auditorium an "ultra-modern" structure and noted how its form had attracted significant interest from the local artist community.[24] The signature structure was later described as an "oval dome" auditorium, with a seating capacity of twelve hundred people, an "acoustical masterpiece."[25] These descriptions indicate the multifaceted appreciation of thin shell concrete—as economical, expressive, and spatial.

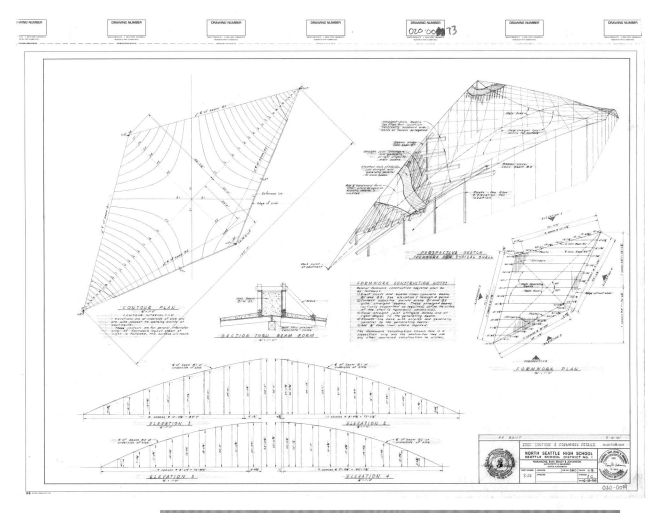

Ingraham High School
Auditorium, Seattle.
(*above*) Roof contour
and formwork details,
drawings dated 1958.
(*right*) Aerial view of
formwork and scaffold-
ing, 1959.

Ingraham High School Auditorium, Seattle. (*top*) Ground view of formwork and scaffolding, 1959. (*bottom*) Shell complete, 1959.

Ingraham High School Auditorium, Seattle. Completed auditorium, 1959.

As with the Mercer Island Multipurpose Room, however, the formwork for the Ingraham auditorium was extensive. While it used only straight-line pieces of tubular scaffolding and dimension lumber, the building required a massive assembly of supports under the shell, filling the volume of the building it was creating. The maze of the temporary structure also had to be exact in its geometry, tightly specified to achieve the proper form of the shell. Christiansen had recognized the usefulness of the segmental hyperbolic paraboloid as a compositional design strategy, but he had not yet employed it as an efficient construction method. Regardless of these challenges, Christiansen was finally finding traction for hyperbolic paraboloids in the Northwest.

NATIONAL SCENE

Directly aligning with Christiansen's work, 1958 was a breakout year for hyperbolic paraboloids nationwide. From 1955 to 1957, national publications had shown hyperbolic paraboloid shells in small-scale applications, such as gas station canopies, reservoir covers, and restaurant buildings.[26] By 1958, architects and engineers were designing thin shell concrete in larger, more architecturally significant structures across the United States. In August 1958, *Architectural Forum* noted: "Thin shell construction, an advanced structural method of enclosing space, is reaching a critical stage of development in the U.S. It is not only finally accepted and thriving, but increasingly fashionable, a dangerous stage for any new engineering or art form. The quip is that h.p. no longer stands for horsepower but for hyperbolic paraboloid, a swooping shell form that fills many architectural dreams these days. This may be a slight exaggeration, but, with some large pockets of resistance remaining, the shell has arrived."[27]

The article described how hyperbolic paraboloid thin shell structures were designed as not only warehouses but also churches and schools and in signature, long-span conditions. Anton Tedesko had branched out from the barrel vault as well, designing the largest hyperbolic paraboloid structure to date—May D&F's entrance canopy in Denver, with the architect I. M. Pei (1957–59).[28] Other large-scale structures were popping up across the country. The architectural firm Harrison & Abromovitz, with engineers from the office of Ammann & Whitney, were working on the University of Illinois Assembly Hall (1957–63)—a radial, folded-plate shell spanning over 400 feet in diameter at Christiansen's alma mater.

Also in 1958, Candela published a reflective article in *Architectural Record*.[29] In it he praised the hyperbolic paraboloid as the easiest and most practical shape to give a shell. However, Candela described a disconnect that was growing between architects and engineers in understanding the geometrical form. Resisting the rampant use of the form, he stated: "Too often . . . the projects using this shape are quite unbuildable. This is because of the prevalent conviction that absolutely any structure is possible. Nothing could be falser than this. . . . Its use in a few buildings of rare and unusual shape has helped to encourage a kind of collective fury to design extravagant and pseudo-structural forms. This naturally places the poor people whose mission is to calculate such structures in a very unhappy position."[30]

Candela hoped to add a structural rigor to the form-based excitement surrounding hyperbolic paraboloids. The distance between designer and calculator that Candela described was exactly the divide that Christiansen was negotiating on a daily basis in the Pacific Northwest.

Responding to this international interest and discussion surrounding thin shell concrete, the legendary Spanish engineer Eduardo Torroja saw the need for a community of architects, engineers, and builders who were interested in shell construction. In September 1959, Torroja convened a colloquium on shell structures in Madrid and proposed the founding of the International Association of Shell Structures (IASS).[31] Torroja saw the organization as a venue for both academics and professionals to share their research directions and new shell design ideas. The organization would hold annual meetings, publishing a regular bulletin of activities, and it continues to serve as a centralized institution for disseminating shell design activities.

This widespread use of concrete shells only strengthened Christiansen's conviction that the forms of thin shell concrete could be both sophisticated and expressive architectural structures. He would continue creating structures through the segmental assembly of warped panels, but also work to refine their construction method. To continue to build with thin shell concrete, Christiansen would have to leverage the repetitive geometry of the design into a repetitive (and cost-effective) construction procedure.

UNIVERSITY OF WASHINGTON PEDESTRIAN BRIDGES

As a consulting structural engineer, working in an engineering (not architectural) office, Christiansen also found design opportunities outside of the strictly building market—in transportation infrastructure. In 1956, President Dwight Eisenhower signed the Federal-Aid Highway Act authorizing construction of an interstate highway system that would connect the major urban centers of the United States with a high-speed road network. In the coming years, Seattle would soon be increasingly connected to the east by Interstate 405 and north-south by Interstate 5. With the prolific expansion of the private automobile, Seattle had a growing need for roads, bridges, and parking lots.

Major institutions in Seattle began to plan for the influx of cars, through the construction of new thoroughfares, the widening and rerouting of existing roadways, and the expansion of parking facilities. In November 1957, the University of Washington hired Worthington & Skilling to study and design parking and access facilities east of Montlake Boulevard.[32] The firm rarely worked independently of architects, but with extensive site work and civil engineering required, Skilling was able to obtain a prime contract from the university.

The design aimed to link different corridors of traffic while also providing pedestrian access to campus, calling for several vehicular and pedestrian overpasses. In the preliminary design, Skilling described the engineering approach to the design of overpasses and structures of the interchange: "A

University of Washington–City of Seattle Montlake Interchange, Preliminary Design, Seattle. Model photograph, ca. 1957.

minimum use of materials has been used to obtain the lightest appearing structures consistent with safety, rigidity and economy. This is achieved by the use of solid slab cross sections with the cross sections varied to suit the stress distribution. The entire, full depth of these slabs is maintained uncracked, and hence usable, by the application of post-tensioned, prestress forces. Practical considerations require that these slab shapes be economically formed. As a result, although the form surfaces are warped, they are ruled surfaces which may be formed of straight lumber."[33]

In his description, Skilling was clearly applying Christiansen's growing expertise in hyperbolic paraboloid geometry and concrete prestressing to provide an economical solution to the structural problems of bridges. Even though it was a transportation project (not an architectural project), Skilling still promoted Christiansen's design as "the lightest appearing structure," achieved through an economical, minimal use of material.

By leveraging the new expertise of the firm, Skilling was also providing a new opportunity for Christiansen to design bridges for the first time. Without an architect on the project, these bridges show Christiansen's own structural aesthetic. Bridges had fascinated Christiansen ever since he saw the bridges of Robert Maillart in college, and he now had the opportunity to design ones of his own. As part of the preliminary design, Christiansen designed several bridges that shared attributes with Maillart's. Christiansen's bridges had a low, slightly curving profile on top, while the depth of the cross section varied at different locations in the span. The bridges had slender, tapering columns and slabs with rounded corners where they intersected.

TYPICAL SECTION
THROUGH DECK

TYPICAL SECTION
THROUGH PIER

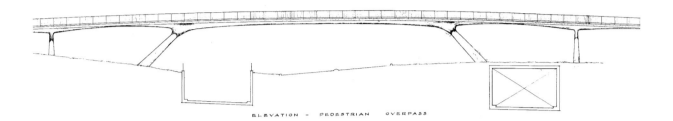

ELEVATION — PEDESTRIAN OVERPASS

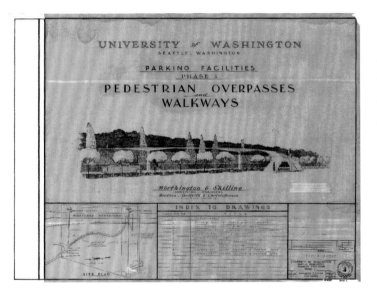

(*top*) University of Washington–City of Seattle Montlake Interchange, Preliminary Design, Seattle. Sketch of typical section and elevation of pedestrian overpass, ca. 1957.

(*bottom*) University of Washington Parking Facilities, Pedestrian Overpasses and Walkways. Title sheet, 1958.

In these designs, Christiansen combined his knowledge of the hyperbolic paraboloid with Maillart's minimalist aesthetic, while pushing the geometry into a structurally impressive, long-span condition.

In 1957, the University of Washington Regents decided to build a large, 7,000-space surface parking lot east of Montlake Boulevard, with two pedestrian bridges spanning over the roadway west to the campus. The pedestrian bridges had to connect a distance of 138 feet between the lower parking lot and the upper campus walkway. Christiansen modified his preliminary designs and specified a single bridge design to use in two different locations. The bridge was symmetrical about its center, with two vertical columns on either side of the roadway (78 feet apart) and two extensions on either side (each 30 feet long) connecting to a spiral staircase on one side and a pedestrian path on the other. Using the hyperbolic paraboloid, Christiansen warped the underside surface of the deck, giving the bridge a triangular section of varying depth. Keeping the width of the deck constant, the bridge deck was thickest at the column locations and thinner toward the middle (where forces were low). This geometry allowed the top edge of the bridge deck to stay thin—8 inches thick—throughout the entire length of the bridge.

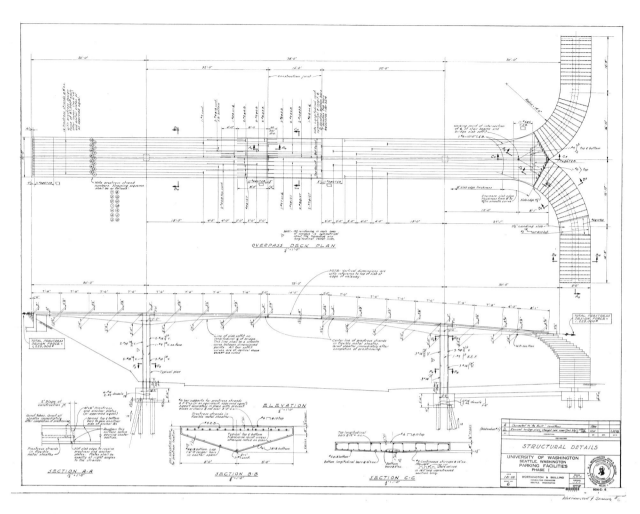

(*above*) University of Washington Parking Facilities, Pedestrian Overpasses and Walkways. Structural details, 1958.

(*left*) University of Washington Pedestrian Bridges, Seattle. Under construction, 1958.

Christiansen was pleased with the design. In his pursuit of thinness, he considered that the underside of the bridge was nearly always in shadow, and the thickness there would not be as visible as it would on the edge. Christiansen also prestressed the concrete—again, a cutting-edge use of this technique—to increase capacity and ensure that the bridge would not show cracks over time.[34]

Christiansen's bridge design also lent itself to a straightforward construction process that allowed for multiple uses of the same formwork. As the bridge design was symmetrical, the contractor was required to build only half the formwork—with half the bridge capable of standing on its own as a cantilever structure. Once cured, the formwork was lowered and moved to the other side of the bridge. This also meant that only half of the traffic on Montlake Boulevard had to be stopped at a time. After one side was cast, the other lane was shut down and the first opened, and the second half of the bridge was formed and poured. With two identical bridges, the same formwork and sequence was repeated for the second bridge.

The parking plan had called for two pedestrian bridges, but the university still had reservations about the cost of Christiansen's bridge design. As a result, the university sent two separate configurations out to bid: bid A, for two overpasses and walkways, and bid B, for one overpass and walkway only.

University of Washington Pedestrian Bridges, Seattle. Completed.

Once opened, the bids reflected the repetitive efficiency of Christiansen's design: the cost of a second pedestrian bridge was minimal. The repetitive nature of his concrete forming system essentially enabled the construction of two bridges for the price of one. In December 1957, the University of Washington Regents approved the construction of two pedestrian overpasses, and both bridges were completed by June 1958.

On this project, Christiansen effectively used a repetitive formwork system for his hyperbolic paraboloid designs, just as he had done with his earlier cylindrical shells. In addition, Christiansen began to understand the benefit of a design that could be self-supporting during construction. With the freedom to independently control the geometry of the structure, Christiansen found that he could maximize the design to be both functional and cost-effective. Christiansen would continue to prioritize the reuse of concrete formwork in a sequenced construction process, all culminating in a carefully designed structure, in his future work.

KING COUNTY INTERNATIONAL AIRPORT HANGAR

Bassetti & Morse was particularly interested in Christiansen's thin shell forms and in the architectural potential of a design based on the reuse of concrete formwork. Following the Mercer Island Multipurpose Room, Bassetti & Morse designed the Men's Physical Education Building at Western Washington University (1959) with a gabled configuration of thin shells to create a long-span space. The firm also designed Mercercrest Junior High School on Mercer Island (1960) as a series of pitched-roof shell forms enclosing classrooms and a gymnasium. While the *Seattle Times* remarked on the school's appearance as "an eye-catcher, with its tent-like roof structure," the architects celebrated the practical benefits: "The Architect [Bassetti] says that the hyperbolic paraboloid umbrellas have many advantages. They offer low initial cost (because one form for the concrete may be used over and over, they offer low maintenance cost and, because they are concrete, the fire insurance rates are lower)."[35]

In 1961, Christiansen worked with Bassetti & Morse on the construction of an airplane hangar at Boeing Field, also known as the King County International Airport, located roughly five miles south of downtown Seattle.[36] In designing Hangar 9, as it was called, Christiansen was interested in simultaneously pushing the structural, aesthetic, and constructive possibilities of hyperbolic paraboloid shells of concrete.

The hangar was smaller than the B-52 hangar at Moses Lake, but still had to incorporate multiple aircraft at once. The hangar was sized to hold five Fokker F-27 passenger planes or two Boeing 727 aircraft, requiring a span of 240 feet, with a height of 68 feet and a length of 155 feet. Following the success of the University of Washington bridges and other projects with Bassetti, Christiansen knew the benefit of designing a building that could be built in segments. If he could design the hangar to be self-supporting as the different segments were cast, the same formwork could be reused many times.

In addition, Christiansen continued to experiment with the hyperbolic paraboloid geometry. Rather than begin from the idea of the warped plane (as he had on other projects), Christiansen started with an alternate (and equally valid) definition of the geometry as the intersection of parabolas of opposite curvature. Starting with a parabolic arch crossing the 240-foot span, Christiansen traced another parabola (25 feet wide) to create a segment of the roof surface. This created a concave (sagging), shell shape in between arches. Keeping his efficient forming methods in mind, Christiansen knew that running straight lines between different intervals on each spanning arch could form this roof surface. These arching segments, with an opposite curvature, gave the profile of the hangar an undulating form, uniquely suited to resist forces of gravity.[37] Working as a freestanding arch, these individual segments could be built one at a time and required only minimal connection between them. The joints between the segments permitted slender skylights to allow natural light into the space. Christiansen described each segment on the structural drawings: "One complete 25' by 240' bay of roof shell with its

King County International Airport, Hangar 9. Under construction, showing movable formwork in place, 1962.

King County International
Airport, Hangar 9.
(*above*) Internal view, 1962.
(*left*) Complete, external
view, 1962.

supporting abutments, columns and ties is a self-supporting structure (with
respect to vertical loads). A single form may be used six times to pour the
building. . . . Forms may be tubular metal scaffolding, or braced timber or
steel falsework mounted on tracks or rollers. When lowered 4' 1¼" minimum, the form may be moved to an adjacent bay."[38]

For the hangar construction, a special fabricated trailer with hydraulic
lifting mechanisms was used to lower false work sections, move them into
adjacent bays, and jack them into place. Thus, six uses of the form were

achieved, and the structural cost was less than a competing design in structural steel.[39] The system could be used to create a vaulted roof of any length, with a shell only 3.5 inches thick.

As the shell segments reached the ground, they had to distribute gravity and lateral forces to the foundation. Christiansen designed two diagonal columns to support each segment, as trussed abutment legs, delivering force to the foundation. With the long-span space defining the entire building, the input from the architects of Bassetti & Morse was minimal. The geometry of Christiansen's structural design became the primary aesthetic. The hangar was inexpensive, functional, and architecturally striking.

MAURY PROCTOR

With a systematic reuse of forms, Christiansen's thin shell concrete designs were becoming less expensive to build, and he saw a great potential in thin shell concrete as a means to create inexpensive, utilitarian space on a widespread basis. Just as Christiansen had found architectural clients fascinated by the warped geometry, he now needed the right construction partner to expand the use. He needed an enterprising builder who could build thin shell concrete formwork and market the thin shell system, and he soon found just the right person.

Maury Proctor (1919–2008) was a local entrepreneur with an interest in construction. Born in the Northwest, Proctor had served as a pilot over the Pacific Ocean during World War II, and after returning to the Pacific Northwest, he became active in the aviation community. Interested in new business ventures, Proctor dabbled in a variety of professional endeavors, including developing an airfield on the east side of Lake Washington.[40]

Surrounded by rampant construction, Proctor sensed an economic opportunity and began building homes in the Seattle area. He first worked with established builders and gained experience and insight into how homes were built. Despite having no formal education in engineering or construction, Proctor proved to be extremely inventive and industrious, always thinking of ways to save time and money.

In 1954, he began to develop a series of products aimed at reducing the time and cost of residential home construction, through a business named Proctor Products.[41] His first product was the Proctor Wall Jack—a tool used by residential builders to lift wood-frame walls from the deck to a vertical position—and he moved on to a set of reusable metal forms. Rather than building each form out of wood and throwing it away after use, Proctor designed a durable form that could be used over and over again—primarily for the casting of foundations. Proctor also modified the wall jack into another system for casting concrete walls on the ground and tilting them up into place, capable of being operated by only two men. Within the booming construction industry, Proctor found a large market for his concrete forming system.

To grow his forming system into a marketable, proprietary product, Proctor needed to consult a structural engineer. He needed an engineering stamp to verify that the forms were strong enough to hold the weight of the concrete before it cured. Though it required only a simple calculation, Proctor needed to talk to an engineer.

One day in 1959, Proctor called Jack Christiansen. Precisely how he selected Christiansen, among all the engineers practicing in the region, is unknown. Christiansen easily completed the formwork calculation, but instead of simply hanging up the phone, Christiansen and Proctor continued to discuss concrete work in the region. Christiansen was intrigued by the forming system that Proctor had started and could see the potential benefit of this system applied to thin shell concrete. Proctor could sense a new commercial endeavor.

In addition to residential construction, industrial warehouse construction was also on the rise—a market that Proctor was eager to tap into. Proctor had come up with a means of building foundations and walls out of concrete, but he did not have a system for building roofs. At the time, roofing systems were typically framed of heavy timber beams, light wood trusses, or light gage steel beams and deck. Proctor and Christiansen both knew that concrete (in terms of material costs) was less expensive than all of these options, and if they could develop a low-cost forming solution for overhead roofs, they just might have an effective system for a mass-production, all-concrete building.

In 1959, as a test of the reusable system, the two collaborated on an office building for the publication *Pacific Architect and Builder*, designed by its editor, architect A. O. Bumgardner.[42] This building used hyperbolic paraboloids in a gabled configuration, requiring tie rods to counteract outward forces through interior space. A similar system was used on the smaller, Shannon & Wilson office building, designed in 1959 as well, with NBBJ. While the tie rods were not ideal (as they interrupted the interior space), the reuse of formwork on these projects indicated the wider potential.

Christiansen and Proctor soon turned to the freestanding, hyperbolic paraboloid umbrella—four shell segments arranged around a central column, which Christiansen had first used in Wenatchee. With a refinement of the proportions, these umbrellas could become both roof and column—

covering 30 feet by 30 feet in plan—with concrete less than 4 inches thick. Christiansen and Proctor set to work on a reusable formwork system for these umbrellas. They came up with a steel frame topped with fiberglass-coated plywood that was a durable casting surface capable of enduring hundreds of castings. Each form had two sides that could be brought together to form the roof surface and then lowered once the concrete had cured. The forms fit on the back of a standard highway trailer, making transportation easy. With hydraulic lifting mechanisms, just a handful of workers were needed to set the forms in place. Together, Christiansen and Proctor refined all aspects of the construction process and, as a result, could drastically lower the cost of an all-concrete industrial building. Christiansen credited this forming system with "revolutionizing thin shell construction in the Northwest."[43]

Formwork for hyperbolic paraboloid shell, ca. 1960. (*above*) On truck for delivery to construction site. (*right*) Formwork in place, ready for casting.

(*left*) Jack Christiansen on top of hyperbolic paraboloid umbrella, ca. 1960

(*below*) United Control Corporation, Redmond, WA. Architectural diagram showing layout of hyperbolic paraboloid umbrellas, 1960.

Christiansen and Proctor found their first large-scale project for using the formwork system in early 1960—a warehouse for the United Control Corporation, in the Overlake Industrial Park in Redmond, with architect Paul Kirk (1914–95). At the time, Kirk was becoming one of the most well-known modern architects in the Puget Sound region.[44] Kirk was interested in experimenting with different materials in his buildings, moving away from the brick, steel, and glass of the International Style in favor of more locally sourced materials.[45] An August 1962 article in *Architectural Forum*

identified three "threads" that unified Kirk's approach to modern architecture: "clarity, suitability, and restraint."[46] Predisposed to the thinness and minimal use of material that shell construction afforded, Kirk recognized that reinforced concrete allowed for forms and details that lent themselves to architectural expression. Concrete could also be designed to span long distances and offered new spatial conditions as well. In designing a utilitarian space with tight budgetary restrictions, Kirk became interested in Proctor and Christiansen's new concrete forming system and embraced the opportunity to try something new.

The United Control Corporation, founded by three former Boeing employees in 1949, focused on manufacturing thermal switches (a temperature-activated control) for various aerospace applications. With expanding production, the company needed an expansive warehouse that could be easily reconfigured as production changed and yet also have tight environmental control. Kirk, Christiansen, and Proctor came up with a building that was a simple composition of sixty-two freestanding umbrellas (arranged in a roughly rectangular pattern), enclosed by insulating "sandwich" panels (rigid foam insulation in between layers of concrete), all resting on a simple foundation system. In this building, Christiansen could prove that thin shell concrete was both an expressive and a practical, cost-effective means of making signature structures.

The approach was novel in the Pacific Northwest. In 1960, the *Seattle Times* wrote that "'parabs' could be described as concrete structures resembling giant umbrellas that got caught in a high wind. Or a group of bonbon dishes. 'Parabs' is shoptalk for hyperbolic paraboloid structures, which are being used for the United Control Corp. building going up on a 30-acre site in Overlake Industrial Park east of Seattle. The building is one of the first on the Pacific coast to make extensive use of 'parabs.'"[47]

The article continued to describe how, while hyperbolic paraboloid form meant "more strength with less concrete," the required formwork had previously made it too expensive for widespread use. On this project, Proctor and Christiansen's mobile forming system had drastically reduced construction costs. The general contractor on the project was Cawdrey & Vemo—the same company that built Christiansen's first shell project at the Green Lake Pool. With experience building that form piece by piece, Nels V. Nelson, superintendent of the Overlake project, echoed Christiansen in calling this forming system "revolutionary."[48]

The shells were cast with a tight, looping construction schedule. On day one, the shells were stripped from their previous location and moved to the new location. Reinforcing bars were laid. On day two, the concrete was poured and allowed to set for three days. On the sixth day after the forms were erected, they were removed and taken to the next location.

In addition to the speedy, cost-effective construction method, the project was celebrated as a significant architectural work. The August 1962 article on Kirk in *Architectural Forum* prominently featured the warehouse, describing it as a "graceful," "sculptural" building.[49]

SHELL FORMS INC.

Following the success of the United Control project, Proctor decided to set up a formal business venture to pursue more building projects—called Shell Forms Inc. With the forming system able to be used on multiple jobs, the cost of labor for each job was severely reduced and the material cost of concrete was minimal. He soon found that he could create an all-concrete building

Promotional brochure for Shell Forms Inc., n.d. (*left*) Cover. (*right*) Interior page.

Shell Forms Inc. warehouse structure, unknown location, n.d.

that was competitive with wood and light gage steel for warehouse construction, with additional insurance benefits as a fireproof material.

Proctor produced promotional material that advertised "safe . . . functional . . . beautiful" all-concrete buildings for commercial and industrial use: "A Shell Form Building is the only kind that provides 'Umbrella Coverage' for your total business investment. It is Permanent, Safe, Maintenance-Free, Expandable, a Good Investment, Low Cost, and Attractive."[50] The brochure extolled the virtues of a Shell Forms all-concrete building, with reduced fire risks and insurance costs, with increased utility and open space—all made possible by the modular construction formwork system.

For Christiansen, the engineering design was nearly identical for each warehouse and required only minimal oversight, leaving Proctor free to market the building system as widely as possible. Between 1960 and 1977, this technology would be used in eastern and western Washington to build over sixty warehouses and production facilities—potato warehouses, factory buildings, and shopping centers. Proctor aggressively promoted the system and used the formwork on educational projects and public works, including Cedar Valley Elementary School for the Edmonds School District (1962) and the Edmonds Civic Center (1961), with Dan F. Miller, and the Tri-Cities Airport (1967), with Pence & Stanley Associates.

After Kirk's experience with hyperbolic paraboloid shells on the Overlake project, he began to incorporate the warped panels of thin concrete into many of his designs. On some occasions he would use Proctor's forming system for repetitive economy; other times he would work with Christiansen for a more custom shell design.

In 1961, Kirk used hyperbolic paraboloid shells for the design of a small church on a challenging site. The Church of the Good Shepherd wanted to build a 144-seat chapel on a steep hillside lot in Bellevue, a suburb of Seattle.[51] To keep foundation work to a minimum, while retaining the wide-open views of the surrounding terrain, Kirk and Christiansen designed a stacked floor and a roof of shell construction, completely supported by a central pier planted in the hillside. With just one column, the shell cantilevered in every direction—extending toward the open view—and was connected back to the street level by an elevated bridge.

The simple architectural design concept concealed the structural challenge. The floor construction was a 3-foot concrete shell made up by a number of hyperbolic paraboloids fitted together and cantilevering 58 feet from its supporting pier.[52] The pier consisted of a cluster of three smaller columns, coupled together. With an unbalanced cantilever, Christiansen had to rely on the elevated bridge to act as a counterweight, developing large forces in the edges of the shell. The only way to handle the large forces that this configuration produced was to incorporate post-tensioning into the edges of the shell and the bridge structure, supported by a post-tensioned girder cantilevered 72 feet from the pier. This unique geometry obviously did not match Proctor's standardized form, and the entire formwork had to be built from scratch.

To save money for the church, Christiansen proposed that the roof be a plywood shell of identical geometry to that of the floor. While he had never designed a wood shell before, Christiansen was confident that he could get shell behavior out of a wood assembly. In a sense, instead of building wooden formwork and casting concrete on top of it, the formwork could just become the structure itself.

Church of the Good Shepherd, Bellevue, WA. Model view, 1961.

Church of the Good Shepherd, Bellevue, WA. Nearly complete, view from below, 1962.

Christiansen designed the shell as two layers of 0.5-inch plywood nailed to a grid work of dimension lumber members, spaced approximately 4 feet apart.

Thanks to Christiansen's structural ingenuity and confidence with the hyperbolic paraboloid form, Kirk could realize a signature chapel design and the church could keep costs low.[53] In February 1961, the *Seattle Times* reported that "the design, by the architectural firm of Kirk, Wallace, McKinley & Associates, will feature one of the first church usages of a singularly supported 'hyperbolic paraboloid' base which will support an elevated 3,000 square foot chapel."[54] The church was completed in 1962 and dedicated on February 14, 1963.

Christiansen presented his work on the Church of the Good Shepherd at the 1962 World Conference on Shell Structures in San Francisco, organized jointly by the University of California, Berkeley; the IASS (founded by Torroja in 1959); and the federal Building Research Advisory Board. This conference showcased the expanding global interest in shell design, gathering architects and engineers from Mexico (Félix Candela), Japan (Yoshikatsu Tsuboi), India (Joseph Allen Stein), Germany (Frei Otto), France (Nicolas Esquillan), and Poland (Wacław Zalewski). Christiansen felt right at home with this organization.

More locally, Paul Kirk became a major proponent of the hyperbolic paraboloid shell—using the form in many different projects. In 1960, Kirk partnered with Christiansen on the Wood Products Research House in Bellevue—an experimental residence roofed by a series of wood hyperbolic paraboloid sections.[55] The details for the project were published in *Progressive Architecture*.[56] In 1966, Kirk also designed the Mercer Island Beach Club, a two-story clubhouse of stacked hyperbolic paraboloids, built using the repetitive Shell Forms Inc. formwork system.[57]

Symbolic of Christiansen's total acceptance of the hyperbolic paraboloid geometry, in 1964 he built his own residence using the Shell Forms Inc. system. For a site on the western side of Bainbridge Island, facing the Olympic Mountains, Christiansen served as both architect and engineer for his family home. He designed a six-bedroom, three-bathroom house that consisted

(*left*) Christiansen Residence, Bainbridge Island, WA. Completed 1964, contemporary view.

(*below*) Mercer Island Beach Club, Mercer Island, WA. Completed view, 1966.

of just three 24-foot-by-24-foot and four 30-foot-by-30-foot hyperbolic paraboloid shells. These umbrellas were used in a combination of single-story and double-story (stacked) configurations. With this arrangement, the entire house was supported on just five columns, minimizing the foundation work on the steeply sloping site. Christiansen borrowed a set of architectural details from Kirk to help with the interior wood finishes, window details, and shading devices. One hyperbolic paraboloid umbrella supported an exterior deck, and another covered an exterior carport.

The August 1961 issue of *Architectural Forum* featured young architects primed to "shape American building in the decade ahead."[58] Also highlighted was new talent among engineers (under the age of forty-five), among them Myron Goldsmith (Skidmore, Owings & Merrill), Edward Cohen (Ammann & Whitney), and Stefan J. Medwadowski (University of California, Berkeley). Christiansen (aged thirty-four at the time) was included in this group. The article asked (by survey) about several topics concerning the "future of structures." Regarding reinforced concrete, Christiansen stated: "It is very important that we do not lose sight of the plastic nature and possibilities of concrete and find ourselves merely producing the same straight posts and beams in concrete that have been used in wood and steel (and concrete)."[59] Continuing, he said: "There is an attitude prevalent, among a good many architects and engineers, to the effect that, in the USA, we are unable to do refined structural designs because our high labor costs make them uneconomical. This, I believe, is a fallacy. . . . The actual reduction of material quantities does save money."[60]

With the hyperbolic paraboloid, Christiansen had found his new system for thin shell design. While he would not completely abandon cylindrical shells in the future, his designs were few and far between. When designed to be self-supporting throughout construction, the hyperbolic paraboloid geometry was much more adaptable than the barrel-vaulted form, capable of being used in an even wider range of architectural projects. With this geometry, Christiansen began to have an impact on the modern architecture of the Northwest and would guide his shell forms for decades to come.

CHAPTER 4

AN INTERNATIONAL SHOWCASE

The 1962 Seattle World's Fair

CHRISTIANSEN'S APPROACH TO thin shell concrete was transitioning and maturing at a time when Seattle was changing as well. The 1962 Seattle World's Fair would launch the city into a new level of national consciousness, while creating fairgrounds that would become a permanent civic center for the growing city. Christiansen's designs in thin shell concrete—demonstrated through three separate projects, with three different architects—would play a central role in the architecture of the fair and showcase his work as a fully realized medium of the modern architecture in the Pacific Northwest.

PLANS FOR THE FAIR

Economic prosperity and population growth brought new civic aspirations and a new opportunity to showcase the city and the region on a national scale. Plans for a Seattle World's Fair had been discussed as early as 1955, initially intended for 1959 as a fiftieth anniversary of the 1909 Alaska-Yukon-Pacific Exposition.[1] Eventually settling on the summer of 1962, the governor, Arthur B. Langlie, appointed a World's Fair Commission (chaired by local businessman and hotelier Eddie Carlson) to organize the fair as a onetime event but also create a more permanent resource for the expanding city.[2] Named the

Century 21 Exposition (a.k.a. Seattle World's Fair), Seattle. Aerial view, 1962.

Century 21 Exposition, its buildings and fairgrounds were intended to point toward the future—an exciting display of technological innovation, but also the foundation for the future civic center for the city of Seattle.

The fair proclaimed itself "America's Space Age World's Fair," and the expanding city of Seattle was an appropriate setting.[3] The Boeing Company's jet engine–powered 707, released in 1958, was transforming air travel nationwide, making destinations (like Seattle) more accessible to the eastern parts of the country. Meanwhile, national ventures into outer space were taking on more international significance. In response to the launch of the Russian satellite *Sputnik* in 1957, President John F. Kennedy delivered a speech to a joint session of Congress on May 25, 1961, calling for an ambitious space exploration program. Despite the rising Cold War hostilities and increasing conflict in Vietnam, the president projected a sense of optimism about the future, and the promotional material for the Century 21 fair followed suit.

Never before, perhaps, has world civilization been so absorbed in its own future. The rockets that man sends overhead are marking paths which he will follow. He seeks to free himself from the bounds of Earth. It is the aim of the Century 21 Exposition to portray this new era . . . to preview the ways man will work and play and live in the year two thousand. In this respect, Century 21 will differ from past expositions, offering not a review of man's progress but an insight into his future, a portrait of life in the twenty-first century—"Man in the Space Age."[4]

To plan a fair that met these lofty design intentions, the World's Fair Commission created the Design Standards Advisory Board and, in August 1958, named Paul Thiry (1904–93)—another significant modern architect in the Northwest—as the supervising architect for work on the fairgrounds. Thiry had gained a national reputation for introducing a European modernism to the Pacific Northwest through a series of residential homes, while also serving on several regional planning boards.[5]

As principal architect for the fair, Thiry was charged with creating "an exciting and forward-looking exposition whose buildings and spaces could have all the drama and spectacle usually connoted by fairs, and at the fair's end, a city center for cultural, sport and other community events."[6] Thiry prepared the site plan for the fair—laying out new buildings and incorporating a small number of existing buildings into a unified plan. Looking toward the future, the architecture of the fair was intended to "exemplify the finest of contemporary design and embody ideas, concepts and materials which may prevail in the 21st century."[7] The steel-framed Seattle Space Needle (designed by the John Graham Company with Victor Steinbrueck and structural engineer John Minasian), with its swooping, curved legs and flying saucer–like top, would soon rise as the signature structure of the fair.[8]

Jack Christiansen did not actively pursue World's Fair projects, rather the projects found him. By the early 1960s, his shell designs had infiltrated the modern architectural culture of Seattle and Christiansen was becoming a favorite collaborator of many of the significant modernist architects in the region. Acknowledging the versatility of his design ability, architects began looking to Christiansen to contribute to the aesthetic appearance, as well as the structural design, of their projects. With the Shell Forms Inc. construction method, thin shell concrete was also proving to be a practical and economical total-building system, competitive with other alternatives like steel and engineered wood.

Christiansen's engineering firm was also expanding. In 1960, to recognize the contributions of Helge Helle and Joe Jackson, the engineering firm Worthington & Skilling expanded its name to Worthington, Skilling, Helle & Jackson. Following his impressive work at the Seattle World's Fair, Christiansen would be made the fifth, and youngest, partner in 1962.

The commissions to design individual buildings within Thiry's overall plan were awarded to many different architects in the region. As a consulting engineer, Christiansen was simply working closely with the architects who retained these commissions. In the course of his daily design work, Christiansen soon found himself collaborating on three different projects on the fairgrounds, with three different groups of architects: the US Science Pavilion with Minoru Yamasaki and Naramore, Bain, Brady & Johanson (NBBJ); the International Pavilion with the Spokane firm Walker & McGough; and the Civic Center Exhibition Hall with Paul Kirk (partner at Kirk, Wallace, McKinley & Associates).[9]

These three new projects were in addition to an earlier building on the fairgrounds that was retained and converted for a new use during the fair. Christiansen had designed the barrel-vaulted Nile Temple headquarters (with architect Samuel G. Morrison)

Original Nile Temple (Century 21 Club for 1962 Seattle World's Fair). Under construction, 1956.

in 1956. With its undulating roof profile, the building's appearance fit with the expressive character of the new fair buildings and was converted to the headquarters for the Century 21 Club—a social organization of fair boosters. The building was an artifact of Christiansen's earlier design method (cylindrical shells), but its reuse (rather than demolition) demonstrates how thin shell concrete was a part of the architecture of the fairgrounds from the beginning.

Each one of Christiansen's new projects required a different level of collaboration with the architects involved and resulted in distinctly different designs.

The federally funded US Science Pavilion was one of the largest complexes on the fairgrounds. The fair's promotional material described the pavilion as "a large, ultra-modern building designed to reflect the concept of its contents—the world's scientific vistas" and to express "the role of science in the Space Age."[10] The pavilion would contain five separate exhibits: *The House of Science, The Development of Science, US-Boeing Spacearium, The Methods of Science*, and *The Horizons of Science*. An additional space enclosed *The Junior Laboratory of Science*. Sited on the south end of the fairgrounds, on the corner of Second Avenue North and Denny Way, the complex was intended to express and display the promising technological future of the United States enabled by science—a government showpiece at the core of the fair's cultural mission.

In 1959, the federal government awarded the design of the US Science Pavilion to one of the most nationally prominent architects of the time, Minoru Yamasaki (1912–86).[11] Yamasaki was born in Seattle and earned a bachelor's degree in architecture at the University of Washington in 1934. After graduation, he moved to New York City, working for Shreve, Lamb & Harmon, and then in 1945 moved to Detroit, where the firm of Smith, Hinchman & Grylls recruited Yamasaki to be its head designer. In 1949, Yamasaki started his own firm with two partners, George Hellmuth and Joseph Leinweber, and in 1955, the firm reorganized as Yamasaki & Associates. Yamasaki was quickly emerging as one of the most promising modern architects of the time, with a perspective that sought to humanize the hard lines of the International Style. He merged decorative elements and ornamentation with cutting-edge technology, in a desire to create a sense of serenity and peace in his designs. This interest had already led him to thin shell concrete and experience working with some of the most well-known engineers in the United States.

In 1951, Yamasaki began the design of the Lambert–St. Louis Airport (1951–56). One of the first postwar air terminals, the airport was Yamasaki's first major public building, but also his first venture into thin shell concrete as a design medium: a terminal space enclosed by four groin-vaulted shells. Anton Tedesko, the same engineer Christiansen had encountered in Chicago, served as structural engineer for the project. Consulting with Yamasaki, Tedesko helped define the profile, thickness, and stiffening ribs required for the vaulted shells. Though the building was celebrated in publications for its integration of thin shell structure, Yamasaki expressed regret over the design. Later, in an oral history interview, he said: "There are several things

Lambert–St. Louis Airport. Designed by Minoru Yamasaki and Anton Tedesko. Completed 1956, contemporary view.

wrong with the building. I don't like the way it comes down to the ground at the back. I realize now that we should have expressed the strong corners of the building in the rear facade, also in the front; but it's less conspicuous in the front since the ground is at the first-floor level. But at the rear where it is three stories high it is very conspicuous and makes me very unhappy."[12]

Yamasaki wanted to push the architectural potential of shell design further, and he continued to experiment with the medium. In 1958, he designed the American Concrete Institute (ACI) headquarters, working with the engineering firm Ammann & Whitney. The modest office building had an expressive, thin, folded-plate roof surrounded by a curtain wall of glass and decorative concrete block. Yamasaki was pleased with the result, less as a finished product than as a set in an evolution of experimentation. Yamasaki was also interested in exploring texture and finish in concrete, primarily through precasting (casting concrete elements off-site and transporting them to the job location for assembly into a building form).[13] He would continue many of these directions (thin concrete, texture, and precasting) at the Seattle World's Fair.

During the late 1950s, Yamasaki was returning to the Northwest regularly.[14] In 1957, he was named to the Design Standards Advisory Board for the Civic Center and the Seattle World's Fair, alongside Thiry and others.[15] In January 1958, Yamasaki returned to his alma mater to serve on the University of Washington Regents' Architectural Advisory Commission.[16] And on January 22, 1960, the US government announced that Yamasaki had been named the architect for the federal Hall of Science.[17]

To execute the project in Seattle, Yamasaki partnered with the architects at NBBJ. Yamasaki was a classmate of Perry Johanson at the University of Washington, and the two remained close. In need of a good engineer to execute a technologically advanced building, Johanson introduced Yamasaki to John Skilling and the engineers at Worthington, Skilling, Helle & Jackson. Yamasaki had already worked with some of the most prominent engineers in the nation in Tedesko and the firm of Ammann & Whitney—the leading experts in thin shell concrete. At Johanson's insistence, Yamasaki became familiar with Christiansen's work in thin shell concrete and was particularly impressed by his use of prestressing to reduce the thickness of his shell structures. Yamasaki agreed to hire Worthington, Skilling, Helle & Jackson as a consulting engineering firm.

Yamasaki and Skilling very quickly struck up a close personal and professional relationship, developing a shared creative perspective on the united forms of architecture and structure, most evident in tall buildings. In the years that followed the fair, they worked together on many other buildings across the country—most notably the World Trade Center towers in New York City (1964–73), in association with Leslie E. Robertson (1928–).[18] As a marketer and salesman, Skilling was eager to respond to the design ideas of Yamasaki, promising structural forms at the edge of possibility and driven to make nearly any form possible. In making these promises, Skilling relied on the technical expertise of other engineers at Worthington & Skilling—including Jack Christiansen. While Skilling maintained a relationship with Yamasaki, Christiansen would become the primary structural designer for the US Science Pavilion.

Christiansen's expertise would be put to good use. Yamasaki wanted to continue his expressive exploration of concrete and push the boundaries of what concrete could do. As Christiansen knew, the prestressing of concrete was transforming the material and sectional possibilities, making concrete structures lighter than ever. Yamasaki, however, was not interested specifically in shell structures for the Seattle fair. Despite the national excitement, his own previous experience had made him reconsider the architectural reception of shells.

Yamasaki was interested more in using his architectural design to create a peaceful setting. He wanted to harness the advancing technology and use it to create a human-centered experience. In his evolving perspective on modern architecture, Yamasaki expressed his aim to create buildings to "love and touch."[19] He wanted his buildings to have a sensitive, tactile quality. He was interested in combining the high-quality texture and finish of precast concrete, with the thinness of shell construction, in the creation of a serene environment.

Yamasaki was also influenced by his experience as a visitor at previous world's fairs in other cities—where each architect was "vying to be more spectacular than the next, with chaotic results."[20] As a result, Yamasaki chose to design not a singular building but a central courtyard space surrounded by six separate buildings. The visitor would enter a somewhat enclosed space on a series of elevated platforms, with soaring arched towers overhead and fountains and pools below. In describing his design, Yamasaki said: "We wanted visitors to be intrigued as they first see the five towers of the pavilion—and then the visual surprise of pools and fountains."[21] The courtyard was intended to be an oasis in a rapidly growing Seattle.

The building facades that faced the interior of the complex would shape the pristine environment of this courtyard. Yamasaki wanted these building walls to be as thin and delicate as possible, revealing interior spaces and walkways through openings between columns. At the same time, Yamasaki wanted the buildings to have no interior columns, leaving wide-open floor plans for exhibits of different sizes. To minimize the number of elements and attain this refined aesthetic, the facade of each building would have to become the structure. As a result, each building became a strikingly simple assembly of thin, precast concrete elements: exterior bearing walls supporting long-span, prestressed concrete girders overhead. The project relied on the pattern, detail, and finish of the exterior walls to create Yamasaki's serene environment.

Yamasaki developed a repetitive pattern of columns and ribs to be expressed on the surrounding wall surfaces and worked with Christiansen to make them structural as well. The walls of each building were divided into 5-foot-wide panels. Each panel was only 3 inches thick across its width, but had a thicker portion running down the center (6 inches wide, 1 foot 3 inches thick) that acted as a load-bearing column. Each one of these columns aligned with a long-span roof or floor beam behind it—with a flat, bearing corbel or haunch extending 8 inches on the backside. Between these columns, curving ribs (4 inches wide, 6 inches thick) traced out a pointed arch–like geometry. The ribs created their arching form only when multiple panels

US Science Pavilion, Seattle. Precast wall panels lifted to form bearing walls of buildings surrounding courtyard, 1961.

were assembled side by side to create the building elevation. The flat wall surface in between the columns was solid in some places, open in others— allowing for colonnade-like walkways and windows. To achieve the desired thinness, Christiansen would have to use Yamasaki's pattern as the beginning of his structure design—placing the reinforcing steel and prestressing cables within the pattern of ribs, with little flexibility to alter the overall form.

The walls of the buildings not only shaped the courtyard experience but also supported the long-span roof and floor beams of the building behind. With a T shape, these prestressed concrete beams varied between 20 inches and 38 inches deep, but kept a constant 5-foot width, aligned with the width of each wall panel. The end walls were enclosed by 7-foot-by-6-inch-wide panels, with a similar central column. These end panels were non-load-bearing, with a slightly scalloped curvature, distinguishing them from the load-bearing facade panels.

This precise design would not be easy to fabricate. These panels were both the facade of each building and the structure, requiring careful detailing and finishing but also a careful monitoring of their strength. Described by the fabricator as "thin-shell wall sections," each panel needed to maintain tight tolerances to ensure that Yamasaki's patterns were executed precisely and the reinforcing was put in the right place.[22] This level of detail and precision required a precast concrete approach. In describing the motivations for using precast, the fabricator explained that "the quality and beauty desired and the extremely critical design of the long, thin-shell wall sections made the degree of control that can only be achieved in plant work a necessity."[23]

The wall panels were cast in fiberglass forms with a structural steel support. The precast company, Associated Sand and Gravel, designed custom prestressing equipment to match the tight tolerance and exact rib geometry. It experimented with different mixes of concrete, using a facing mix of crushed white quartz and a backup mix of high-strength concrete. The 364 bearing wall panels and 131 end panels were cast at a plant in the nearby city of Everett and trucked to the fairgrounds for erection between May and October 1961.[24] The result was a finely detailed finish, with a lightness of form.

US Science Pavilion, Seattle. Openings in precast concrete bearing walls allowing passageways and light, 1961.

In the center of the building complex, Yamasaki designed a series of five delicate towers, 101 feet tall, that marked the entrance to the central courtyard. Continuing his interest in lightness and thinness, Yamasaki designed pointed arches filled with a thin lattice. Each tower was composed of four prestressed, precast concrete columns topped with connecting arches that rose higher than the surrounding buildings. The columns were roughly U-shaped in section, with outside dimensions of roughly 2 feet by 2 feet. As they approached the lattice dome, the columns branched into three arches—one connecting to the peak of the dome and the other two branching to the adjacent columns.

Christiansen defined the curvature of these arches in the same manner as he had for his earlier thin shell structures. He defined an equal spacing between the columns and a series of vertical measurements that approximated a parabolic curve. The arches between the columns warped outward across their arc to align with the dimensions and orientation of the adjacent column. His ability to rationalize complex geometry into simple, buildable terms was once again essential.

US Science Pavilion, Seattle. Overhead arches under construction, 1961.

The spaces between these arches were infilled with smaller precast elements—only 6 inches deep by 8 inches wide—in a triangular pattern. To simplify their fabrication, each one of the cross members and diagonal members were straight in their horizontal projects and only curved in their vertical projection. Therefore, the plan of the tower lattice was a simple pattern of triangles.

Yamasaki included a light fixture, suspended between the columns, 47 feet above the slabs below. While likely not fulfilling a structural role, the fixtures were located near the spring point of the top arches. At this location, they appeared to work as a tension tie, resolving any outward thrust generated by the overhead dome. Christiansen detailed these fixtures on the structural drawings, using a 0.375-inch-thick, corrosion-resistant plate steel assembly embedded in each precast column. Four 0.5-inch-diameter steel rods extended to the center point between all four columns, where a globe-like light fixture was supported.

When Christiansen first saw the tower design, knowing Yamasaki's interest in thinness, he began to try to make the towers as light as possible. Christiansen reduced the size of all the pieces to be as small as they could be, which included questioning the need for some of the lattice pieces that Yamasaki had designed to connect the arches together. Christiansen recalled: "As I was drawing the towers, I was taking out all the members that we didn't need. Or ones that I didn't think we needed. Yama was walking down the aisles of the office, when he came to my desk he stopped, he looked at it, and he said: 'That's not right. Put those back in.' So I put them back in."[25]

Indeed, Christiansen's structural drawings for these arches show several erasures and changes to the lattice pattern. This design approach was different from other jobs. Christiansen was used to collaborating with architects, helping shape the overall appearance of buildings, even taking the lead on the external form, but it was clear that Yamasaki had a different approach to collaboration, with less interest in input from structural engineers on the overall form of his architecture. Still, Christiansen recognized and appreciated Yamasaki's design talent and his interest in structural form: "Yama did the shape of things, but to his credit, he did have a strong sense of structure." Later, in 1963, after Yamasaki had collaborated with Worthington & Skilling on several projects, *Architectural Record* recognized his focus in its article "Structure Plays Leading Role in Latest Yamasaki Designs."[26]

With the prescribed latticework and the finely detailed precast work, these towers put the thinness of Yamasaki and Christiansen's mutual design on dramatic display. As a result, the towers seemed impossibly light, in a stunningly advanced use of prestressed concrete. *Architectural Record* headlined an

US Science Pavilion, Seattle. Overhead arches nearly complete, pavilion buildings in the background, 1961.

article "Soaring Ribbed Vaults to Dominate Yamasaki's Design for Seattle Fair."[27] Some claimed that the towers recalled the structural forms of Gothic architecture (e.g., flying buttresses), and Yamasaki responded: "While the form of the arch may be Gothic, I have never seen towers like these in any Gothic architecture."[28]

Yamasaki admired Gothic cathedrals for their structurally derived forms. In 1959, he remarked: "I like the Gothic arch, I'll be frank to admit—but it has a valid reason for being."[29] But Gothic cathedrals also had a "consistency" to their design that Yamasaki sought to emulate, yet he was not interested in re-creating historical forms. "If there is delight in the buildings of the past because they have this richness, for one thing, this enjoyment of shape and shadow, this silhouette, this variety of shapes, why should we in our age of technology . . . why should we limit ourselves to a rectangular architecture? This makes no sense at all. . . . The only thing that we have to be careful about is to have restraint and to have sensitive and serene buildings."[30]

Christiansen, Skilling, and Robertson would play a key role in Yamasaki's future work. In his previous work in concrete, Yamasaki had not achieved the level of structural refinement he desired. Christiansen exceeded Yamasaki's demands, making the towers and tracery thinner than Yamasaki had imagined. Yamasaki admitted: "That's the first time an engineer did me one better."[31]

Yamasaki was so pleased with Christiansen's design work, and the thinness he achieved, that he allowed Christiansen to display his own structural aesthetic in one location. Upon entry to the pavilion complex, visitors arrived at a series of elevated platforms—concrete slabs supported by the same columns of the towers that rose overhead. Rather than support the 40-foot-by-40-foot-square slabs at their corners, Yamasaki wanted them supported at the midpoint on each of the four sides. This configuration made the corners of the platforms cantilever into the courtyard, contributing to a sense that the slabs were floating.

This arrangement created a structural challenge for Christiansen—needing to make the slabs lightweight but strong enough so the corners didn't deflect downward excessively. The slab had to be thick enough to resist the bending moments created in the slab, but thin enough to keep its self-weight low. Freed from Yamasaki's design control, Christiansen chose to explore a new, but related, structural idea.

An avid reader and always interested in exploring a new idea, Christiansen recalled Pier Luigi Nervi's ribbed slab work, which he first encountered in his architectural history courses. The Gatti Wool Factory, completed in 1953, showed Nervi's experimentation with photoelasticity and isostatics, aligning an under-slab rib structure with the lines of maximum and minimum principal moments.[32]

Christiansen admired this pattern, and the arrangement of ribs that reflected the flow of forces, but knew that with the fair's tight schedule he did not have time to develop an isostatic analysis procedure of his own. As a result, he approximated Nervi's system through an easily definable geometrical pattern that he knew was rooted in the same structural principles as Nervi's. Considering a larger scale (40 feet between columns, compared to 16 feet at the Gatti factory), a different configuration (columns at the midpoint of the slab, compared to columns at the corners), and different loading requirements (lighter pedestrian loads, compared to heavier factory loads), Christiansen drew up his own pattern of ribs.

US Science Pavilion, Seattle. Elevated entrance slabs, under construction, showing pattern of ribs and post-tensioning cables, 1961.

Christiansen designed essentially two sets of intersecting ribs emerging from each column support. Each slab was divided into four identical quadrants, each 20 feet by 20 feet. Each quadrant also maintained a diagonal line of symmetry, from the cantilevered corner to the center point of the slab. As a result, only one-eighth of each slab required detailing. A solid, 3-foot-thick section filled a radius out to 4 feet 2 inches from each column. From there, circumferential ribs (1 foot wide, 3 feet deep)—like ripples in a pond—occurred at 6-, 8-, 10-, and 12-foot radius locations, still centered around each column. Another set of eight ribs emerged radially from each column support, equally spaced and angled 11.25 degrees from each other. These ribs crossed the circumferential ribs at right angles, diverging further apart toward the diagonal line of symmetry in the slab.

Christiansen's pattern, inspired by Nervi, also recognized the practicalities of construction. The defined radii of the circumferential ribs were simple to measure in the field, and thin, flexible boards easily matched the specified curvature. A constant thickness and depth of rib made the formwork simple and thick enough to contain something Nervi's design did not—curved post-tensioning cables. These cables, embedded in the concrete and tensioned after concrete was poured, would add to the bending capacity of the slab

US Science Pavilion, Seattle. (*top*) Elevated entrance slabs nearly complete, 1961. (*bottom*) Underside of elevated entrance slabs revealing Christiansen's pattern of ribs, 1962.

US Science Pavilion, Seattle. (*left*) Complete, 1962. (*above*) Aerial view, 1962.

where it was needed the most. A thin flat slab on top provided the walking surface, making the pattern of ribs visible only from the underside of the entry platforms.

Unlike his experience on other thin concrete projects, Yamasaki was thrilled with the results. The *Seattle Times* reported that Yamasaki and Johanson were "tickled to death" by the design of the US Science Pavilion: "[It was] even more exciting than we thought it would be."[33] In its use of thin precast concrete, the building complex had also fulfilled the objective of showcasing the latest technology. The architectural assessments of the US Science Pavilion praised the modern technological methods used to create the complex.[34] Yamasaki and Johanson remarked: "For a science pavilion to show the progress of mankind, technologically we should be way ahead in construction. We think the building represents some real technological achievements, with prestressed, precast concrete techniques."[35] In this project, Christiansen had extended his expertise in shell construction to accommodate Yamasaki's design, further demonstrating his skills as a creative structural engineer. The US Science Pavilion played an instrumental role in Yamasaki's winning the commission for the World Trade Center towers in the years that followed, and, known today as the Pacific Science Center, remains a landmark in contemporary Seattle.[36]

WORLD OF COMMERCE AND INDUSTRY, INTERNATIONAL PAVILION, WITH WALKER & MCGOUGH

While the US Science Pavilion was an exercise in working within Yamasaki's design concepts, other smaller projects at the fair afforded Christiansen more widespread design freedom. For a structure intended to house a collection of international exhibits for the World of Commerce and Industry, called the International Pavilion, Christiansen took a more prominent design lead position. Though he worked alongside the architects of Walker & McGough, Christiansen's structural design would define the overall form of the pavilion.

Walker & McGough of Spokane was awarded the contract for the World of Commerce and Industry in February 1960.[37] Led by Bruce Morris Walker, the firm had previously designed residential homes and a women's residence hall for Washington State University, in Pullman, featured in the Western Section of *Architectural Record*.[38] It would later design Padelford Hall on the University of Washington campus.

The International Pavilion was situated on the northern edge of the fairgrounds, just south of Mercer Street. Through the design of three separate buildings, the pavilion was required to house a variety of exhibits from around the world. The first building was to contain exhibits on the United Nations, Africa, Thailand, and the Philippines, and the second, exhibits on India and South Korea. A third building contained exhibits on San Marino, Ecuador, the city of Berlin, and the Peace Corps.

For the first two buildings (with the largest exhibits), Walker wanted to create a large overhead roof structure to act as a canopy that could unify and contain the many smaller structures and exhibits underneath. Walker was already working with Christiansen on the design for the Washington Corrections Center in Shelton—along with Bassetti & Morse and the New Orleans architectural firm Curtis & Davis—using hyperbolic paraboloid umbrellas as the primary structure (see chapter 5).

The architects and Christiansen decided to again use the hyperbolic paraboloid form, except in a completely new configuration. For this overhead structure, Christiansen designed a modified, more expressive version of the freestanding umbrella. While the typical umbrella consisted of four, square panels arranged around a central column, Christiansen instead designed six wedge-shaped panels to complete an umbrella form. Geometrically, this simple exercise replaced four 90-degree corners with six 60-degree corners, around a central column. This variation in the mathematics of the shell dimensions and proportions created a drastically different effect.

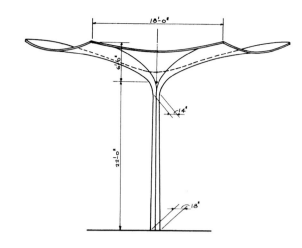

The warped geometry of the individual umbrella began to take on new characteristics. The umbrella form was also now hexagonal (not square), which meant that the arrangement of umbrellas into a continuous roof surface would force the position of the columns to alternate on the site. Matching the geometry of the new shell segments, Christiansen designed a central, 22-foot-tall, fluted column that gradually transitioned into the shell above. Christiansen also sculpted the edges of the hyperbolic paraboloid panel to create an undulating edge, resembling the petals of a flower.

International Pavilion, Seattle. Umbrella geometry, 1961.

Walker and Christiansen arranged fifty-two of these freestanding umbrellas in a compact, hexagonal pattern to cover the exhibits below. When multiple umbrellas were arranged to form a continuous roof surface, the underside of the structure appeared to flow as "an organic appearing shell structure."[39]

Despite this complex geometry, Christiansen could still define a simple, rational construction sequence—afforded by the hyperbolic paraboloid. Christiansen described the precise method in the construction documents, using simple dimensional measurements and readily available materials.

International Pavilion, Seattle. Model view, 1961.

With their increasing experience together, Christiansen recommended Maury Proctor to build the custom formwork for the job, similar to the repetitive forming system for Shell Forms Inc. Proctor built large frames of aluminum, welded them together, and then topped them with a fiberglass-coated plywood for a durable, yet smooth, finish. Only four sets of forms were used to cast all fifty-two shells—as each form was used thirteen times—once again drastically reducing the cost of the structure.[40] A fast-setting, high-strength concrete was used to allow the rapid casting, curing, and form removal, as well as a thickness of only 1.5 inches.[41] With this efficiency, the construction of the pavilion was smooth. Proctor noted: "It was originally anticipated that 17 weeks would be required to pour the 52 shells based on a 7 day cycle for each set of forms. The job was actually completed in 13 weeks due to an improved form cycle that became as low as four days under optimum conditions."[42]

Walker designed a variety of enclosures for the different exhibits, ranging from colorful fiberglass infill panels to lattice wood. A folded-plate retaining wall contained the north and west perimeter of the complex.

The rationality of Christiansen's design generated a striking aesthetic. The roof was said to resemble "rigid umbrellas or sails."[43] An architect's guidebook for the fair described the shells as "Oriental" in appearance: "Located

International Pavilion, Seattle. Under construction, 1961.

International Pavilion, Seattle.
Completed view at night, 1962.

on the northwest corner of the exposition site, the inside-out umbrellas with
their colorful fiberglass panels present an exciting boundary."[44] As a group,
the umbrellas created a flowing, undulating texture visible from underneath.
By simply manipulating the hyperbolic paraboloid geometry and reassem-
bling panels, Christiansen created a new, yet still rational, form with striking
architectural effect.

CIVIC CENTER EXHIBITION HALL, WITH KIRK,
WALLACE, McKINLEY & ASSOCIATES

Christiansen's next project at the fair became an exercise in using thin shell
concrete in a more subdued manner, while still accommodating a long-span
condition. Along the northern boundary of the fair along Mercer Street,
further east from the International Pavilion, the World's Fair Commission
authorized a permanent complex of buildings that would serve as a major
arts center for Seattle.[45] The arts buildings at the Seattle fair were specifically
intended to balance the science and technology emphasis of other exhibits,
to create a well-rounded cultural experience.

This complex included the reconfiguration of the Civic Auditorium into
the new Opera House (by architects James Chiarelli and B. Marcus Priteca,
of Priteca & Chiarelli) and the Exhibition Hall and Playhouse (by Kirk,

Wallace, McKinley & Associates). All these buildings were connected by a 30-foot-high colonnade, visually making them part of a singular complex. Worthington, Skilling, Helle & Jackson provided the structural engineering for both projects. Leslie E. Robertson worked with Priteca & Chiarelli, and Jack Christiansen (as he had done before) worked directly with the architects at Kirk, Wallace, McKinley & Associates.

As the lead designer, Paul Kirk was interested in a more restrained aesthetic at the fair, in direct contrast to the flamboyance of other buildings. *Architectural Forum* wrote: "Kirk has deliberately placed culture in a quiet setting," bringing a sense of calm to the fair.[46] Though no less technologically advanced than other fair buildings, the design of the Exhibition Hall and Playhouse was an exercise in formal restraint.

Kirk designed the Playhouse and Exhibition Hall as an ensemble—two buildings connected by covered walkways with a landscaped courtyard in between. The Playhouse contained an 800-seat theater and lobby, while the Exhibition Hall (also known as the Fine Arts Pavilion) was intended to accommodate a wide variety of uses.

The walkways were composed of 30-foot-tall concrete columns supporting a flat-slab roof. The columns of the walkways were extremely slender, and cruciform in section, with multiple corners and shadows visible from different directions. The roof, strikingly flat and thin, had only modest coffering in between the column locations. The coffered pattern responded somewhat to the structural forces in the flat slab much like Christiansen's platforms for the US Science Pavilion, but instead of many ribs tracing lines of support, the slab simply had a regular grid of rounded square-shaped indentations. This design provided a simpler, quieter rhythm of shapes, giving modest texture to an otherwise flat surface.

This column and ceiling design extended from the walkways into the Playhouse, shaping the spaces of the lobby and covering the Playhouse theater. With the Playhouse's stable, consistent program demands, a regular grid of columns did not disrupt its operation. *Architectural Forum* described the Playhouse as a "sober, quiet structure," appropriate for the "dignity" of the theater.[47]

For the Exhibition Hall, a different program drove a different structural solution. The Exhibition Hall was expected to host banquets, art shows, exhibits, and civic functions.[48] Such flexibility of program required a flexibility of space, preferring minimal interruption from internal columns. As a result, Kirk and Christiansen laid out a perimeter of concrete columns on the rectangular plan 140 feet wide and 220 feet long. They then needed a structural system to span the 140-foot width—a fairly long span—but did not want a particularly expressive building profile. They wanted to retain the quiet serenity of the arts buildings and create a similar design to the flat-roofed Playhouse nearby. The arches and vaults of this scale that Christiansen had designed earlier (for airplane hangars and auditoriums) had all risen significantly above their supports.

To match Kirk's desire for quietness and serenity, Christiansen had to develop a long-span concrete system, with the simplest expression possible. While the span was too large for a flat-slab system, folding the slab into corrugations gave the structure depth without increasing thickness. This folded-plate structure did not have the curvature of shell structures, but remained efficient in its use of material by creating a structural depth through folding. By using the distance between the top and the bottom of the fold, Christiansen was able to span the space in a low-profile manner.

Christiansen designed eleven folded-plate shells, each 20 feet wide and spanning 140 feet between exterior columns, 40 feet above the foundation. He described the design as follows: "The folded plate slab is 4.5" thick throughout, except the bottom slab which is 6" thick. There is a 3'6" wide flat section at the ridge and valley. The overall depth, out to out of the structure, is 6'4.5". This is a span-to-depth ratio of $\frac{1}{22}$, quite shallow for concrete construction."[49] With "folds" only 6 feet deep, the structure kept a relatively low profile overall. To achieve this shallowness, Christiansen designed an elaborate system of post-tensioning, using parabolically draped strands of cables in the inclined slab sections. These cables, once tensioned with jacks at either end, would help lift the center of the span.

Fine Arts Pavilion, Seattle. Architectural model, 1962.

As before, Christiansen designed a repetitive construction method. "The folded plates were cast-in-place, one 20-foot bay at a time. The forms were frames with steel trusses and wood joist stringers with plywood sheathing."[50] Kirk detailed the roof to extend beyond the perimeter columns—with the edge appearing as kinked corrugations visible from the plazas on either side of the building. While expressed on the exterior, the folded profile was still far less dramatic than the other shell structures—at the fair and elsewhere—keeping with Kirk's subdued tone for the arts buildings.

Fine Arts Pavilion, Seattle. (*above*) Roof under construction, 1962. (*right*) Complete, with Space Needle in background, 1962.

This roof design created a signature, undulating profile along the perimeter of the building, in addition to long-span space within the building. Christiansen's use of thin shell concrete in the design was refined and controlled, making it part of the building ensemble, part of a designed complex, rather than a signature element of its own.

These three projects were central structures of the Seattle World's Fair. In the setting of the Seattle fair, Christiansen's thin concrete had taken center stage and showcased itself as a material of the future. Through his earlier work, Christiansen had demonstrated how shaped thin shell forms could be expressive, how prestressing technology allowed concrete beams to be smaller, and how precasting yielded high-quality finishes. But his use of thin shell concrete design at the World's Fair was an acknowledgment of concrete's role as a central medium of design in the era.

The fall 1962 meeting of the ACI, held in Seattle, celebrated the innovative use of concrete at the World's Fair—with presentations on all three projects designed by Christiansen.[51] Harlan Edwards, progress engineer for the fair, claimed:

> Engineering and architecture have combined their talents to create
> in Seattle, history's finest World's Fair. Not only have its structures
> been built economically, in record time, but also with the aid of modern
> design and concrete technology, they have provided permanent,
> column-free areas at a fraction of the usual foundation loadings and
> costs. More than anything else, however, these World's Fair building
> have proved that beauty is not costly—that it only takes competent
> architectural imagination combined with ingenious engineering
> skills, modern materials, and construction know-how—and that ugliness
> per se is not a necessary part of building economy.[52]

The buildings of the World's Fair were widely celebrated at the time. Joseph Gandy, the fair's president, called the buildings "Exhibit A for American architecture in 1962."[53]

Christiansen's structural design ability played a fundamental role in this achievement. Each one of his new projects required a different level of collaboration with the architects involved and resulted in distinctly different designs. Yet not only was Christiansen able to collaborate with architects and develop innovative structural ideas, but his design ideas were central to the success of each one.

Based on the number of projects he was personally involved with and the expressive, aesthetic impact of his structural forms, it can be argued that Jack Christiansen was one of the most important designers of the 1962 Seattle World's Fair. Unified only by Christiansen's engineering, these buildings displayed a remarkable range of structural techniques in service of modern architectural design.

CHAPTER 5

EXPANDING AUDIENCE

Concrete Shells across the United States, 1963–1971

I N THE YEARS that followed the 1962 Seattle World's Fair, Christiansen's career adapted to a very different design landscape as he worked less with architects in the Northwest and more on a national scale. As a favored engineer, Christiansen collaborated with Minoru Yamasaki on several projects around the country, creating hybrid works incorporating both designers' talents and interests. He also expanded a relationship with the architects at the New Orleans firm Curtis & Davis and put the emerging capabilities of computer analysis to use in the design of one of the longest-spanning shell structures in the world. Work with Shell Forms Inc. continued to produce warehouses throughout Washington State, in addition to larger, low-cost municipal structures. Christiansen furthered his interest in long-span bridges and infrastructure works in projects such as the Nalley Valley Viaduct. Overall, Christiansen's design approach had shifted—from experimenting with spatial forms to extending the capabilities of the medium in aesthetic and long-span applications. These design experiences both overlapped and set the stage for the complex Kingdome project to come.

THE 1960S

The 1962 World's Fair was a seminal moment for the Pacific Northwest and for Christiansen's career. The fair presented Seattle as an emerging modern metropolis, both fundamentally changing its national image and expanding the aspirations of its citizens. For Christiansen, the fair showcased his talent as a structural designer, demonstrating his wide range of structural capabilities (precast concrete, concrete shells) and his ability to work with some of the most prominent modern architects of the time (Yamasaki, Kirk, and those at Walker & McGough). At the fair, Christiansen had solidified his reputation as a masterful designer of thin concrete—a talent that could be utilized to fulfill a wide range of architectural intents.

Yet in the years that followed, the national optimism and excitement that the fair created was soon tempered by a series of significant events. President Kennedy was assassinated in November 1963. The United States' involvement in the Vietnam War escalated significantly in 1964. Even as Neil Armstrong landed on the moon in 1969—the culmination of years of government investment in technological advances—rising tensions with the Soviet Union instilled a deep sense of fear in the American public.

Locally, the Seattle economy continued to prosper in the 1960s, but the social and political contexts were changing drastically. Increased public activism and the civil rights movement began to reshape Seattle politics. Civic design projects came under high levels of scrutiny, requiring a longer design and approval process on most building proposals. At the same time, architectural trends were shifting. The minimalist architecture of the late 1950s embraced an expressive thinness—a thinness tied to ideas of refinement and efficiency within architectural modernism, exemplified by thin shell concrete. As the 1960s progressed, this architectural position was gradually replaced with a more complex understanding of buildings and a changing aesthetic palette. Concerns about energy usage, environmental impact, and the creation of public space began to take precedence over structural efficiency.

Yet thin shell concrete had already been firmly established as a competitive, modern building material. Thin shell structures continued to be built around the United States, most significantly in long-span applications. These buildings included University Hall at the University of Virginia, designed by the architects at Baskervill & Son, with the engineers of Severud Associates (1965); the Norfolk Scope Arena, designed by Pier Luigi Nervi (1968–70); and St. Mary's Cathedral in San Francisco, designed by Pietro Belluschi and Nervi (1967–71). The design of concrete shell structures was a significant topic of international debate, with the International Association of Shell Structures (IASS) holding its annual conference in Brussels in 1961 and San Francisco in 1962.

The office environment at Worthington, Skilling, Helle & Jackson was also changing in the 1960s. In 1964, Leslie E. Robertson was elevated to partner alongside Christiansen (who made partner in 1962). The commercial jet enabled the firm to expand nationally, working with Yamasaki and other architects on projects throughout the United States while maintaining its primary office in Seattle. By 1967, both Harold Worthington and Joe Jackson had retired, allowing the firm to reorganize and recognize its current leadership in a name change. The firm was renamed Skilling, Helle, Christiansen & Robertson.[1]

Christiansen transitioned into a leadership role in the office, assuming the position of design lead or partner-in-charge, while delegating analysis and detail work to others in the office. In this arrangement, Christiansen was still able to act as a conceptual collaborator on projects, but, with assistance, was able to increase the number of projects he worked on. For much of the work after the World's Fair, Christiansen no longer drew all the structural details for individual projects by himself, though he continued to guide their design.

Christiansen's work in the 1960s is best understood through these shifting conditions. His expertise in designing thin concrete structures flourished, but his designs would have to change and adapt to a shifting set of architectural priorities. The hyperbolic paraboloid—as a pure structural geometry—would be modified, adapted, and incorporated into more complex building compositions. At the same time, Christiansen designed transportation infrastructure and civic facilities that show that his structural engineering design work (as integrated with architecture) was not just an objective, calculable science but a creative, practical, and collaborative endeavor.

This constellation of complex projects also sets the stage for the project that had the single largest impact on his career: the Seattle Kingdome. These projects were undertaken as the long, complicated Kingdome project unfolded, revealing the source of the central structural concepts and design conditions that Christiansen later applied.

WORK WITH YAMASAKI

Yamasaki had been impressed with Christiansen's work at the Seattle World's Fair. With the success of the US Science Pavilion, Yamasaki recognized Christiansen's ability to integrate structure and architecture—to listen to the architectural intentions (and follow orders) but also contribute ideas and execute finely detailed designs. Christiansen's design work had exceeded the standard set by previous engineering consultants, and Yamasaki would use the engineers at Skilling, Helle, Christiansen & Robertson on many future projects—no matter where in the country the project was.

The two firms would soon work together on the largest project either office had ever retained: the World Trade Center towers in New York City. At Yamasaki's insistence, John Skilling interviewed for the project in 1962—just months after the opening of the Seattle fair—and despite being the only non–New York engineer to audition, Skilling won the job, immediately giving the firm international recognition. In 1964, Robertson opened a New York office to complete the project and give the firm a truly national presence. Skilling and Robertson both came to have close personal connections with Yamasaki, through their work on the World Trade Center. Christiansen was never interested in moving to New York City or in vying with Skilling and Robertson for Yamasaki's personal affection. Christiansen remained content in designing individual buildings, thankful for the opportunities that Skilling and Robertson provided.

Still, Yamasaki never forgot the technical expertise of Christiansen and his ability to design thin concrete structures. As time moved forward, Yamasaki became interested in designing a wide variety of building forms—some dictated by structure, others not. Yamasaki continued to design with precast concrete, but remained skeptical of concrete shells. For Christiansen, working with him was both challenging and rewarding, as Yamasaki continually demanded new structural solutions for a variety of building types and spaces.

Shortly after the fair, Yamasaki designed several buildings for colleges and universities across the country, including Oberlin, Wayne State, Butler, Princeton, and Harvard.[2] All of these projects used an assembly of precast concrete elements that continued the design ideas of the US Science Pavilion. Often rectilinear in plan, these buildings typically had a series of tightly spaced columns, open lower colonnades, and brilliant white finishes. Yamasaki also continued to explore and extend his architecture through structural form in reinforced concrete, and Christiansen was there to contribute to and execute the structural design.

NORTH SHORE CONGREGATION ISRAEL

In 1963, Yamasaki designed North Shore Congregation Israel in Glencoe, Illinois—Yamasaki's first large religious structure. Interested in a more carefully crafted spiritual space, Yamasaki wanted to influence the experience of visitors to the synagogue through a structure that was an interlacing of daylight and solids, executed in concrete. In its structure, it would contain several similarities to Christiansen and Maury Proctor's warehouse designs.

North Shore Congregation Israel, Glencoe, IL. Completed exterior view, 1963.

The sanctuary space was 50 feet high and 80 feet wide by 126 feet long, framed by eight pairs of opposing cast-in-place concrete fan-vault shells. These monolithic shells combined the roof and column structure into a single cast piece. Though not defined by a hyperbolic paraboloid geometry, each concrete shell was identical, making it possible for only two sets of forms to be built and each one used eight times, in sequence, to form the entire concrete roof. By casting two shells at a time—one on each side of the sanctuary space—the structure could be self-supporting during construction. In this design, Christiansen's pragmatic logic of repetitive casting had infiltrated Yamasaki's work, forming these fan vaults in a similar manner to the hyperbolic paraboloid shell warehouses of Shell Forms Inc.

The concrete end walls were also cast in place, while the enclosing sidewall panels were precast—distinguishing them from the fan-vault shells and expressing their nonstructural character. These precast walls, again, were similar in their construction to the shell warehouses of Shell Forms Inc.—as precast, tilt-up walls surrounding a cast-in-place roof and column structure.

But as at the US Science Pavilion, Yamasaki brought a sensitivity to the design of this system. Fully in control of the shape of things, he utilized Christiansen's expertise in thin shell concrete and shaped a new type of space. Also

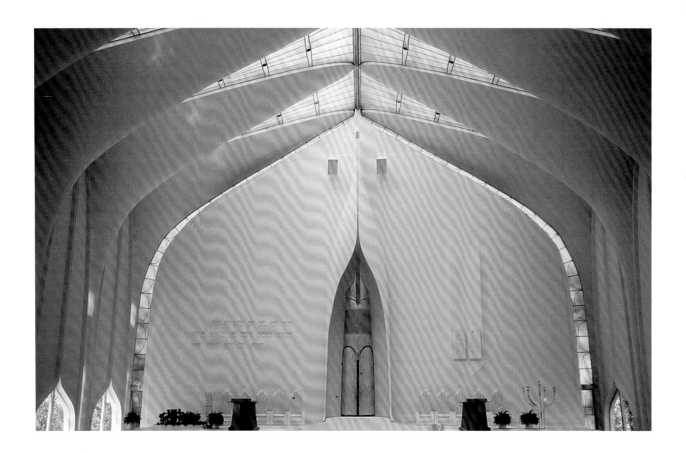

as before, Yamasaki was pleased with the result: "The space gives a sense of uplift, and being able to see the sky and nature outside seems appropriate to a contemporary house of worship."[3] Yamasaki also reinforced the importance of structure to his view of a complete architecture, stating that "'the aesthetic totality of any beautiful thing . . . such as a lovely plant . . . is in its concept, its structure, or its tiniest detail.'"[4]

North Shore Congregation Israel, Glencoe, IL. Completed interior view, 1963.

CARLETON COLLEGE

In 1965, Christiansen designed another project with Yamasaki, the Carleton College Gymnasium in Northfield, Minnesota. For this project, Yamasaki finally had Christiansen design a true, thin shell concrete structure—a groin-vaulted space. Yamasaki had been reluctant to use thin shell concrete in his architecture—perhaps wary of following his early design with Anton Tedesko of the Lambert–St. Louis Airport (also a groin-vaulted space). In *Architectural Record*, Yamasaki commented: "Since shells have been springing up all over the country . . . , we have been hesitant about their use. However,

Carleton College Gymnasium and Swimming Pool, Northfield, MN. Completed exterior view, 1965.

this seemed a natural and appropriate place to use them because of the large spans needed to enclose the pool and the gymnasium."[5]

Working closely with Christiansen, Yamasaki came up with three parabolically shaped groin vaults to cover the long-span spaces—two to cover the swimming pool and one to cover the gymnasium. Each groin vault spanned a space 60 feet wide and 120 feet long. With a column in each of the four corners, the shell geometry traced parabolic arcs along the perimeter, rising roughly 15 feet above the top of each column at its midpoint. These arches then extruded to the center point of the space, creating the overall form from the intersection of two perpendicular, parabolic arches.

As before, Christiansen was able to convert this specific geometrical form into legible directions for the builder. On the structural drawings, he specified a step-by-step procedure for locating the columns and tracing the perimeter arcs through a measurement of equal horizontal spaces and their corresponding vertical elevation. From these defining arcs, the builder had only to "connect opposing points" on each arc with "horizontal stringers" to define the entire shell surface.

Even with this different geometry—aligning with neither his earlier cylindrical vaults nor his hyperbolic paraboloids—Christiansen's sensibility for defining a buildable system came through. He also continued to design for a reuse of formwork. Yamasaki fully adopted Christiansen's reliance on formwork, and in describing the use of these concrete shells in this configuration, Yamasaki explained: "This particular kind of shell was selected because the natatorium is approximately one-half the size of the gymnasium. Hence, we could logically develop the three-time use of the shell, and benefit in terms of economics by reusing the formwork twice."[6]

Christiansen had accommodated Yamasaki's required control over the architectural form, understanding Yamasaki's drive for thinness, and together they produced significant work. *Architectural Record* described the structures as "graceful shapes" that provided "economy, uncluttered space." The article added that "the design of the Men's Gymnasium at Carleton College is an example in architect Minoru Yamasaki's continuing search for 'the beautiful expression of structure.'"[7]

The Carleton project can also be seen as a refinement of Yamasaki's earlier airport design in St. Louis. Given Yamasaki's dissatisfaction with the external, expressed ribs and support conditions, this project provided a second chance to sculpt and detail a vaulted structure. Instead of the upstanding ribs that traced the intersection of the vaults at St. Louis (visible from the exterior), Christiansen placed the ribs on the interior of the shell and gradually tapered the thickness to minimize their visual impact.

Christiansen's influence on Yamasaki is undeniable. Christiansen was enabling Yamasaki to explore new directions in structure that few other structural engineers could provide. Through their shared projects, Christiansen also got the opportunity to define structure in new and changing contexts. These projects extended his range as a structural designer and allowed him to develop new structural strategies.

Carleton College Swimming Pool, Northfield, MN. Completed interior view, 1965.

CHRYSLER STYLING CENTER DOME

In 1968, Yamasaki was hired to design a "state-of-the-art styling and design center" for the Chrysler Corporation at its Highland Park headquarters, outside of Detroit.[8] Yamasaki designed a complex of interconnected buildings, including two body studios and a domed showroom for "viewing clay models of proposed car designs under ideal lighting conditions."[9]

Yamasaki relied on Christiansen for the design of the domed showroom. Needing to enclose a centralized, long-span space (150 feet in diameter), Christiansen returned to the hyperbolic paraboloid geometry he had used earlier. Christiansen had designed centralized spaces with this geometry before (the Mercer Island Multipurpose Room and the Ingraham High School Gymnasium and Auditorium), with comparable spans (115 feet and 160 feet, respectively). Yet both of these shell geometries had low overall heights (32 feet and 28 feet, respectively), creating a shallow slope to the shell. In addition, each had been designed to rest directly on the ground, allowing the foundation abutments to take large amounts of outward thrust with no architectural consequence.

In contrast, Yamasaki designed the Styling Center's dome to be elevated off the ground on 20-foot-tall columns around a circular (rather than a faceted) perimeter. This meant that all outward thrust had to be contained within a perimeter tension ring around the top of the columns. In addition, the dome had to rise an additional 44 feet from the top of the columns to create a massive interior volume.

Facing these new constraints, Christiansen combined techniques he had used on earlier projects. As at Mercer Island and Ingraham, he divided the centralized space into radial segments—but instead of using a small number of segments (seven and three, respectively), he used sixteen. With a larger number of segments, which now resembled wedges, he could more accurately approximate a rounded dome and, potentially, gain more savings in the reuse of formwork. Borrowing from another project, the King County International Airport hangar designed with Bassetti & Morse, Christiansen defined parabolic arcs as the edges of the sixteen segments to reach the top point of the vault and used straight-line formwork to create a hyperbolic paraboloid surface in between. A compression ring at the top united the structural action of the ribs and segments as a whole. An article in the *ACI Journal* described the dome as follows: "The umbrella-shaped structure consists of 16 rib segments and concave sections, formed in hyperbolic paraboloid surfaces. The 16 angled segments with 16 parabolic ribs join on top in a 16-angled cupola. The tension ring, about 20 feet above ground level, supports a 16-angled precast exposed aggregate fascia."[10]

The structure supporting the dome was cast using normal weight concrete, and the dome was made of lightweight concrete. Seeking to take advantage of his design to the fullest, Christiansen proposed two construction sequences

Chrysler Styling Center Dome, Highland Park, MI. Completed exterior view, 1969.

for building the dome. The first sequence was a rather conventional approach of creating all the scaffolding and formwork at once and casting the dome in a single pour. The second was more innovative, with potentially more cost savings. Christiansen described it as "a mechanically removable form system supported on a temporary tower at the center and on the rail around the inside of the tension ring," where only an opposing pair of segment forms would be required to cast the entire dome.[11] These forms could be used to cast two segments at a time on opposite sides of the dome and, resting on each other, be self-supporting on their own. The forms could then be lowered, rotated, and used to cast the remaining segments, a process repeated eight times to cast the entire dome.

On this project, however, the contractor decided that the structure and the mechanism for the second method were too costly and time-consuming for the company's capabilities. The contractor chose the seemingly faster and more economical method of using heavy-duty scaffolding to form and cast the entire dome all at once. The contractor used a rather conventional method of casting monolithic domes—a fifty-hour operation in which lightweight concrete was continuously placed in five-foot rings, starting from the base.

In this project, Christiansen first articulated the idea of a reusable formwork system for the construction of a large-scale dome. Though it went unused, this innovation shows his willingness to expand not only the scale of thin shell concrete but also the cost advantages of form reuse. The Chrysler Styling Center dome suggested the potential of an efficient, hyperbolic paraboloid dome and served as a clear precedent for the significantly larger Kingdome project yet to come.

WORK WITH CURTIS & DAVIS

Christiansen also began to work closely with the New Orleans architectural firm Curtis & Davis—first on projects in Washington State and eventually across the nation. Nathaniel "Buster" Curtis (1917–97) and Arthur Q. Davis (1920–2011) were both graduates of Tulane University. Davis had gone on to get a master's degree from Harvard and then worked in the office of Eero Saarinen, before returning to New Orleans.[12] Curtis and Davis formed a partnership in 1947. Like other architects with whom Christiansen worked, both designers were interested in using new structural technologies and construction techniques in their buildings.

In 1956, the Curtis & Davis firm designed the Louisiana State Penitentiary in Angola using reinforced concrete. The design called for a double arch of concrete to cover the kitchen and dining room facilities and flat slabs over the walkways. These slabs were built using the patented lift-slab construction technique, where slabs were poured on the ground and lifted into place, eliminating the need for scaffolding.[13]

Following Angola, Curtis & Davis was recognized nationally for its expertise in the design of correctional facilities.[14] The firm combined an interest in innovative concrete with a drive to remake prison facilities across the country. In 1961, Washington State began planning a large correctional center near Shelton. The state hired the architectural firm Bassetti & Morse, but brought in Curtis & Davis to consult on the project.[15] Through Bassetti & Morse, the state hired Worthington & Skilling for the structural engineering, and Christiansen came on to the job.[16]

At Shelton, Curtis & Davis had an interest in using the design of the building as a tool for rehabilitation. As Davis explained, "There were no walls around the complex; a unique decorative detention device replaced the old-time use of bars. Heavy emphasis was placed on education, vocation, and physical training. The Washington Corrections Center began producing mostly self-supporting, self-respecting citizens. The architecture became a tool for correction."[17]

Continuing its exploration of materials, Curtis & Davis decided that the structure would be thin shell concrete. Working with Proctor, Christiansen

used the Shell Forms Inc. system to construct over two hundred hyperbolic paraboloid umbrellas as the roof. In four locations, Christiansen designed a gabled configuration of shells for even longer spans. With this innovative use of concrete as a roof, the Shelton facility was called "the most up-to-date prison in the world" and boasted the latest in "security, rehabilitation."[18]

Christiansen's work with concrete at the Shelton facility opened a new architectural relationship with Curtis & Davis. In 1963, the Skilling firm (as Worthington, Skilling, Helle & Jackson was commonly referred to) designed the IBM Building in Pittsburgh with Curtis & Davis, as an innovative steel diagrid with Leslie E. Robertson as the lead engineer. As part of the project, Christiansen designed another concrete pedestrian bridge—with curving ribs of support underneath the wide walkway and crossing the Boulevard of the Allies to a nearby public park. Christiansen also designed a small church, St. Frances Cabrini in New Orleans, with a cantilevered barrel vault, completed in 1963.[19]

Fully aware of Christiansen's talent, Curtis & Davis in 1964 called on Christiansen to help design a massive international trade center in New

Washington Corrections Center, Shelton, WA. Completed exterior view, 1961.

International Exhibition Facility, New Orleans. Completed view, 1967.

Orleans. The exhibition facility was required to cover 250,000 square feet, within which large column-free spaces were needed to accommodate constantly changing exhibits, conventions, shows, and large meetings. Curtis & Davis wanted to mirror the flowing waters of the nearby Mississippi River, with an expressive and "fluid architectural form" both long-span and economical to build.[20] The firm wanted a design to "give the building lightness and individuality."[21] Though it had worked with other structural engineers, after the Shelton project it turned back to Christiansen to design a signature roof in thin shell concrete.

Christiansen came up with a design that was a variation on his earlier barrel vaults to cover a roof area of 420 feet by 452 feet. He designed seven, 60-foot-wide barrel shells, which first spanned 129 feet over the main entrance drive, the entrance vestibules, smaller offices, and meeting rooms and then continued on to span 253 feet over the main exhibition hall. To provide entrance canopies, the roof cantilevered 30 feet on all four sides of the roof, filling out the full dimensions of the site. To increase the fluid nature of the building, the architects and Christiansen curved the barrel vaults in the opposite direction as well—giving the building its signature humpbacked form. Even though this change in geometry gave the building some double curvature to the shells, it was not significant enough to change the underlying structural behavior—that of long-span cylindrical vaults.

The basic shell was only 4.5 inches thick, yet some locations required post-tensioning tendons to handle large amounts of tension. Christiansen placed the tendons within the shell thickness, increasing the thickness to 5.5 inches in these locations. Instead of the rectangular edge beams he had used on earlier cylindrical shells, Christiansen filled in the valley between vaults, creating a solid V-shaped section of concrete. This section, 5.5 feet deep, contained additional post-tensioning and behaved structurally as an edge beam. This roof was supported on three rows of seven columns, spaced at a minimum of 60 feet apart. The columns were diamond-shaped and tapered to flare out to form a capital at the top to merge with the roof shell overhead.

The structural analysis for the roof was initially based on the original methods and tables from ASCE Manual 31. But working at such a large scale, Christiansen performed an additional beam-method analysis for the cylindrical shell to further verify these results. Though these analyses may have been sufficient, Christiansen took the design even further, by writing several computer programs to solve for more exact forces and stresses within each shell. Structural engineers at the Skilling firm had begun using computers for structural analysis in the 1960s. Leslie E. Robertson, working with computer specialist Richard E. Taylor, had performed intense computer analysis for the World Trade Center towers in New York City.[22] The firm had purchased an IBM 1620 Model 2 computer to run multiple structural analysis equations simultaneously through a system of punch cards, a technique showcased at the 1962 World's Fair in Seattle. From that point forward, the firm had fully integrated computer analysis into its design operations. Christiansen learned the programming language of FORTRAN through a class at the University of Washington and quickly determined how useful computer analysis could be in the design of his shell structures.

For the International Exhibition Facility, Christiansen wrote his own programs to calculate the internal forces in the long-span shell system. Still

International Exhibition Facility, New Orleans. Top view of post-tensioning cables draped within the shell, 1966.

directed by his experience using the cylindrical design guide, Christiansen could now get a much more detailed understanding of the longitudinal force, shear force, transverse force, transverse moment, and deflection within the shells.

This analysis, however, still took a lot of time to run on the IBM 1620. The larger the structure to be modeled, the more time the computer needed to run, with analyses sometimes taking days to complete. Christiansen found that the repetition and symmetry he had designed into the structure for the potential reuse of formwork also offered computational advantages. He designed each of the seven vaults to have an identical geometry and thus had only two types of conditions to analyze in the computer: an interior shell and an exterior shell. Not only that, but Christiansen could theoretically split each shell down the center into symmetrical halves and use the computer to analyze only half shells—on the assumption that the other half was working in a similar way. These simplifications in the computation analysis reduced the complexity of the computer model, gave Christiansen a better understanding of the behavior of the structure, and saved valuable computation time.

The construction of the facility was tightly controlled, detailed by Christiansen on the design drawings. He called for the 253-foot span to be cast first and then post-tensioned using large hydraulic jacks. With a careful placement of tendons and construction joints, Christiansen wanted to be sure that

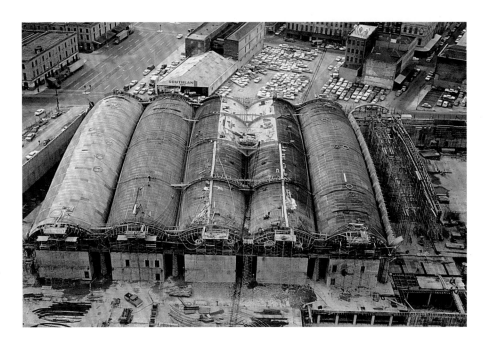

International Exhibition Facility, New Orleans. Segmental casting of long-span barrel vaults, 1966.

the post-tensioning force created a uniformly distributed stress condition. Next, the 139-foot span was cast, followed by the 30-foot cantilever.

The formwork for the cylindrical shells was prefabricated, similar to the formwork created with Proctor. The formwork was built in 20-foot-long-by-60-foot-wide sections, as a series of round wood trusses supporting planks and a 0.375-inch plywood surface. Christiansen assisted in developing a tightly sequenced construction process of forming, placing reinforcing steel and post-tensioning tendons, pouring concrete, and stripping forms. The project was formed and cast between August 1966 and January 1967.

The International Exhibition Facility, known locally as the Rivergate, was considered a major architectural and engineering landmark in New Orleans.[23] It reflected the city's increasing focus on attracting conventions and national tourism and also became a popular local center for special events

International Exhibition Facility, New Orleans. Worker preparing vault formwork for casting, 1966.

International Exhibition Facility, New Orleans. Completed side view, 1967.

and civic activities, including Mardi Gras balls and high school graduations.[24] The building was widely celebrated for its long-span space, but also the sensitivity of the design in reinforced concrete. Architect Arthur Q. Davis recalled: "We spanned 253 feet with slabs that were 5½" thick. Although the walls of the building itself were hammered reinforced concrete, the contrast between the enclosure and the floating roof was unique. Not only was that great swooping roof beautiful; it also allowed for a major porte-cochere and canopy for people arriving and departing the exhibitions that were held in the Rivergate during its heyday. The great open interior spaces could be divided in almost anyway."[25]

Christiansen's contributions to the project are again evident in the praise for the project from the architects. Davis continued: "There is no question that the Rivergate was a significant monument in the city of New Orleans. . . . It was an expression of the sixties. Intended to be an exhibition hall, the Rivergate was a graceful and romantic structure using reinforced concrete in a way that was light and delicate."[26]

Christiansen was internationally and nationally recognized for his design. He presented the design at the 1967 IASS symposium in Mexico City.[27] The building was given an honor award from the Consulting Engineers Council. Christiansen attended the council's national awards competition in May 1969 and presented a poster titled "Post Tensioned Long Span Barrel Shells for the International Exhibition Facility, New Orleans, LA." The display

board proudly showed the results of Christiansen's computer analysis, which were hand-drawn.
A headline in the London-based *Construction
News* declared the Rivergate the "World's Largest
Shell Roof."[28] The journal celebrated the "architectural and structural" features of the building,
demonstrative of the "effectiveness of shell roof
construction." *Engineering News-Record* quoted
a Curtis & Davis spokesman, who claimed: "We
think the roof is considerably different from anything that has ever been built."[29]

Shortly after the Rivergate, Curtis & Davis
acquired its largest project: the design of the
Louisiana Superdome. The Superdome was
part of the era of massive buildings, where cities built huge multipurpose
spaces intended to house as many people as possible, for a variety of different events.[30] Skilling, Helle, Christiansen & Robertson would likely have
managed the structural engineering for this project as well, had the firm not
already engaged in a large-scale stadium project of its own, the Kingdome
in Seattle.

Jack Christiansen (*left*) with
John G. Reutter (*right*) of
the Consulting Engineers
Council, 1969

BRIDGES

Christiansen also continued to design transportation infrastructure. Since
he was not working directly under architects on these projects, these designs
allowed Christiansen even more freedom to explore his own ideas. The Skilling firm had designed several segments of Interstate 5 during its initial construction between 1958 and 1962, with typical box girder overpasses. As a
continuation of this contract, Christiansen got the opportunity in 1963 to
design a signature pedestrian bridge crossing the interstate corridor. The
Shoreline School District (just north of Seattle) had been divided by the
highway construction and needed a pedestrian bridge to connect students
on one side of the interstate to their school on the other side.

The bridge needed to span over both north and south lanes of Interstate 5,
at Northeast 195th Street in Shoreline. For this bridge, Christiansen located
three columns: on either side of the highway and in between the north- and
southbound lanes. Christiansen inclined the columns on either side of the
highway—projecting them over the roadway before they reached the bridge
deck. These sloped columns actually decreased the overall long span of the

Pedestrian Bridge over Interstate 5 at 195th Street, Seattle. Completed view, 1965.

bridge over the roadway, while slightly framing the roadway below. In addition, these columns were tapered, starting wider at the base and narrowing as they approached the upper walkway.

From these support points, Christiansen employed a similar geometry to that of the University of Washington pedestrian bridges. With the double-cantilever method, he was able to build out over the highway below without disrupting traffic. In 1965, the *Seattle Times* described the bridge as "a handsome overcrossing at Northeast 195th St."[31] The article continued: "The pedestrian sky bridge is so graceful and blends so much with the surroundings that, while not unnoticed, it has been mostly taken for granted by most motorists."

Christiansen post-tensioned the concrete deck to make the bridge as thin as possible. The article reported: "Nine tendons of high-tensile steel stressed to 3,500 pounds each were used to permit a silhouette as slender and graceful as a fashion model, at least in the eyes of a bridge builder." The bridge was enclosed by a series of nearly circular, curved pipes along the walkway and stressed, doubly curved wire netting over the top to prevent objects from falling onto the roadway. It also gave pedestrians a wide-open view, while not obstructing the view of passing drivers. The article declared that all these elements combined to create "the graceful architecture of the bridge, in itself a feat of economy."

NALLEY VALLEY VIADUCT

In 1969, Christiansen began design of an elevated segment of Washington State's highway system—his first project to accommodate heavy vehicular loads. Just south of Seattle in Tacoma, the state needed an elevated roadway to connect north-south Interstate 5 with State Route 16, which continued east over the Tacoma Narrows Bridge. The spur, or viaduct, would carry six lanes of traffic across the Nalley Valley—a small depression just west of the interstate that had filled in with various businesses and surface streets.

To bridge the valley, Christiansen designed a unique system of support underneath the roadway. In an attempt to minimize interference with the roadway below, Christiansen identified support points between lanes of traffic, and rather than bringing up single vertical columns, he brought up four, diagonal columns that splayed outward to support the road deck in different locations. This created angled (not vertical) columns that directed their load to a single, larger foundation. Because the four columns balanced each other, each column would still act in direct compression. Attempting to describe these unusual supports, the local media referred to them as "four-columned, inverted tetrapods" and "the first of [their] kind in the state."[32] The inclination of the columns also recalled Christiansen's earlier hangar design at Boeing Field, with Bassetti & Morse. These structurally expressive elements also served a very practical purpose. By having the elevated roadway supported in four points—rather than one—Christiansen could reduce the required span of each beam. A smaller span meant that each beam had to resist less load and could be significantly smaller.[33] The structure deck was cast in place and post-tensioned for additional strength. Underneath, Christiansen designed a pattern of thickened ribs to connect the columns' support points, with rounded coffers in between. The rounded pattern of ribs underneath was to reduce the stress concentrations in the slab and make release of the formwork easier.

Nalley Valley Viaduct, Tacoma. Completed view, 1971.

BUILDING DESIGN & CONSTRUCTION

ARCHITECTS/CONSULTING ENGINEERS/CONTRACTORS   COMMERCIAL/INDUSTRIAL/INSTITUTIONAL

MARCH 1970 *A Cahners Publication*

Cover: Skilling, Helle, Christiansen, Robertson . . . an exclusive BD&C profile

(*left*) Nalley Valley Viaduct, Tacoma. View from underneath, 1971.

(*right*) *Building Design and Construction*, March 1970, featuring John Skilling (*left*), Jack Christiansen (*center*), and Helge Helle (*right*)

Working without an architect, Christiansen was able to make innovative structural decisions without first having the aesthetics of the design approved by anyone else. The inclined supports gave the structure a unique appearance that reflected his own aesthetic as a designer. With this infrastructure project, the state was primarily interested in the practical benefits of the design, and Christiansen accommodated these demands through his design. The district engineer in charge of the project said that "the structure was designed to give it a floating appearance and blend esthetically into its surroundings."[34]

The structural design recalled, as did the slab work at the US Science Pavilion, the design work of Pier Luigi Nervi. Nervi would later design a similar tetrapod column system for the Italian Embassy in Brazil (1971–79), supporting a ribbed floor system above. While it was possible that the two saw each other's work in publications, there was no direct communication between them. This similarity of Christiansen's and Nervi's projects, however, reveals a commonality in their overall approach to structural design.

JOHN B. SKILLING   HELGE J. HELLE   JOHN V. CHRISTIANSEN

LESLIE E. ROBERTSON   ARTHUR J. BARKSHIRE   RICHARD W. CHAUNER

ROBERT O. FOWLER, JR.   KENT R. ROGERS   WM. D. WARD

Promotional material for Skilling, Helle, Christiansen & Robertson, ca. 1975

In March 1970, the publication *Building Design and Construction* profiled Skilling, Helle, Christiansen & Robertson: "In the architectural decade just passed, the most decisive aspect of the overall trend in building forms has been creative assays into bold structural schemes integrated with architect strivings for more convincing architecture. No firm of structural engineers has contributed more to this advance than the firm Skilling, Helle, Christiansen, Robertson."[35]

The article presented the firm's designs in steel through the World Trade Center towers and the US Steel Building in Pittsburgh (led by Robertson and Skilling) and work in concrete by Christiansen's "gracefully thin shelled" Rivergate exhibition hall. The article singled out Christiansen's distinct contributions to the office, as maintaining a "reputation for unusual ingenuity in concepts," where "no project excites him more than a thin-shell or a space frame structure."

The article described the "frontier character" of the firm's work. John Skilling shared aspects of structural engineering that were vital to the firm's success, including using computer analysis to optimize designs to reduce costs, employing methods that conserved materials, and implementing labor-saving processes. The design values that had drawn Christiansen to thin shell concrete had permeated the culture of the firm.

CHAPTER 6

THE GRANDEST SCALE

The Seattle Kingdome, 1963–1976

THE SEATTLE KINGDOME was the largest, most complex project of Jack Christiansen's career. The building took over ten years to go from planning discussions to completed construction. The building shifted sites several times, faced fluctuating civic support, and was under constant budget scrutiny. By the time it was completed in 1976, the Kingdome's final design revealed this long, tumultuous history—a product of the shifting economic, cultural, and political landscape of Seattle.

For many, the Kingdome was the catchall solution to the growing appetite for professional sports in Seattle. The Kingdome was hoped to be an architectural solution to the aspirations of the growing city, where a single large building could contain all the sporting events that Seattle needed to arrive on the national scene. Others saw the Kingdome as an unnecessary yet imposing monument that neglected community impacts and a symbol of the ineffectiveness of county government.

But for Christiansen, the Kingdome was always a design challenge. As an engineer, he saw the stadium as an arrangement of structural elements, whose refined composition could perfectly serve all the functional demands placed on it. In its final design, the stadium was an efficient structural form, capable of serving a wide range of programmatic demands—a model of structural material and cost efficiency. For Christiansen, that was the ultimate goal of any structural design.

In the late 1950s, the importance of professional sports was on the rise throughout the United States, with a particular focus on expansion to the cities of the West Coast. The network of regional leagues for baseball, football, basketball, and other sports gradually dissolved into singular leagues for each sport, operating on the national scale. The new sports landscape of fewer, better teams started a competition between cities for the select franchises of the different leagues. For many, if Seattle was to be considered a significant city in the United States, it had to have professional sports teams, in the most prominent leagues.[1]

Other cities on the West Coast, including San Francisco and Los Angeles, had acquired professional sports in a piecemeal fashion. The Cleveland Rams moved to Los Angeles in 1946, and San Francisco was awarded the expansion franchise that would become the Forty-Niners in 1950. Major League Baseball followed later, with the relocation of the Brooklyn Dodgers and the New York Giants to Los Angeles and San Francisco, respectively, in 1957. Even though San Francisco and Los Angeles were significantly larger cities than Seattle, the arrival of national sports teams made these cities even more prominent and visible to a nationwide audience through televised broadcasts of games. While relatively few homes had televisions before 1947, national broadcasting expanded rapidly in the 1950s.[2]

Aerial view of downtown Seattle and waterfront, 1968

The central issue, often, was the specific venue or stadium that these new professional franchises would play in—with cities promising construction of new facilities. The new teams in Los Angeles and San Francisco first played in existing stadiums, but in the early 1960s, they moved into new venues. Candlestick Park opened in 1960, as the home of the San Francisco Giants, replacing the older San Francisco Seals stadium. Dodger Stadium broke ground in 1959 and opened in 1962. The historic Kezar Stadium near Golden Gate Park was the first home of the San Francisco Forty-Niners; in 1971, the team moved to share a renovated Candlestick Park with the San Francisco Giants. If Seattle were to attract professional sports of this caliber, it would need a comparable professional sports stadium.

In 1960, King County commissioned a stadium feasibility report from the Stanford Research Institute. The report stated that "the Seattle area meets the requirements of Organized baseball and professional football for a Major League franchise, except for the existence of an adequate sports stadium."[3] Following these findings, businessman David L. Cohen asked the Board of King County Commissioners to initiate a $15 million bond issue for a stadium. Cohen had previously approached other municipal bodies, including the Seattle City Council, but was told by Mayor Gordon S. Clinton that the city lacked bonding capacity. Among the first specifications, including the need to seat roughly fifty thousand people, was the requirement for an "all-weather dome"—deemed necessary due to the persistent rains of the Pacific Northwest.[4] Located somewhere between Seattle and Tacoma, the stadium would be a regional resource for the greater Puget Sound area. In a letter to the King County commissioner, Cohen repeated the findings of the report that a stadium was "necessary to attract major-league baseball and professional football to the Northwest."[5]

On November 8, 1960, a $15 million bond issue proposal for an all-weather stadium was narrowly defeated, achieving 48 percent of the vote. The uncertainty of a professional franchise to occupy the stadium, and the exceedingly low price tag of $15 million, led to public skepticism. The estimated cost of the stadium was likely taken directly from the recently completed—yet non-enclosed, open-air—Candlestick Park (designed by John Bolles & Associates), with little consideration of the specific regional conditions.[6]

But after the 1962 World's Fair, public interest in a domed stadium project drastically increased. The fair had fundamentally changed the ambitions of Seattle as a whole. As a prospering city, Seattle was ready to cast off the label of regional center and become a city of national significance.[7] Work was already under way in converting the fairgrounds into a permanent civic center, and momentum behind a permanent facility for professional sports increased as well.

In 1963, the Seattle Chamber of Commerce explored the possibility of a floating stadium in Elliott Bay. Donald N. McDonald & Associates developed a speculative design. In a written statement, the chamber issued its response: "This is thoroughly keeping with the 21st Century outlook and the can-do attitude that has been a part of our community thinking during recent years."[8] The stadium would be supported on floating concrete pontoons, just like the newly opened floating bridges across Lake Washington.

It would be located on the waterfront north of Pier 70 and connected by an extension of the Seattle monorail, built for the fair.[9] A group of Seattle architects and engineers developed preliminary plans for a domed structure on cellular pontoons and justified this location through a combination of demographics and availability. Engineer Brian Maither stated: "We checked a map to locate the center of Seattle's population. But there was no land available in such an optimum spot, so we decided to go out on the water."[10]

STADIUM COMMITTEE REPORT

The floating stadium did not proceed past the proposal stage, but it did help spur the creation of the Washington State Stadium Commission in 1963. The commission was assembled to gather information to be made available to those groups interested in promoting a sports stadium for Seattle and King County. State Representative John L. O'Brien spoke at a sportswriters and sportscasters luncheon in January 1964, reporting on the commission's findings and producing several of its key conclusions. Echoing the "all-weather" demands of the earlier report, the commission indicated that "consideration should be given to a covered stadium, because of Seattle's climate." The commission found that the "estimated cost would be in the neighborhood of 20 to 24 million dollars, and it should be imperative that the stadium be designed for multipurpose use."[11]

HOUSTON ASTRODOME

At the time, with many American cities vying for professional teams, multipurpose stadiums became part of a national conversation. Later that year, in November 1964, the first multipurpose stadium, the Houston Astrodome, would be completed—with a single, steel-framed dome, 710 feet in diameter.[12] It served as the new home first for the Major League Baseball team the Houston Astros in 1965 and then the National Football League (NFL) team the Houston Oilers in 1968. It was called the "Eighth Wonder of the World" and advertised as the "first and only fully air conditioned, enclosed sports arena."[13] With a circular plan, and a dome overhead, the Astrodome was hailed as the future of stadium design, with a price tag of roughly $35 million.

The Astrodome captivated the interest of the Seattle stadium group as both a structural spectacle and a civic resource. The Astrodome contained

Houston Astrodome, Houston. Designed by Hermon Lloyd & W. B. Morgan (with Wilson, Morris, Crain & Anderson consulting) and engineers Roof Structures Inc., Walter P. Moore, and Praeger, Kavanagh & Waterbury. Completed view, 1964.

over 9 acres of land, covered by a network of steel trusses arranged in a domed, lamella configuration. The architectural firm Hermon Lloyd & W. B. Morgan designed the dome, with architects at Wilson, Morris, Crain & Anderson consulting, but engineering concerns with the long-span structure dominated the design process. As the structural engineering of the Astrodome was a significant feat at the time, three different engineering firms participated in its design. Roof Structures Inc. from St. Louis performed the lamella roof design, having designed smaller such structures before. The calculations were then verified by the local Houston firm Walter P. Moore Engineers and by the New York architecture-engineering firm Praeger, Kavanagh & Waterbury.[14] Led by engineer Emil Praeger (1882–1973), the firm had designed Dodger Stadium in Los Angeles (1962) and Shea Stadium in New York City (1964), and Praeger was becoming a national expert in large-scale sports facilities.

The Astrodome seemed to provide the expansive, climate-controlled multipurpose space that Seattle sought, and Houston soon became the model for Seattle to follow.

To spearhead the effort, the state appointed Joseph Gandy to be the chairman of the State Stadium Commission. Gandy had led the 1962 World's Fair effort, and state and local officials were eager to replicate its civic energy and success. In January 1966, the county hired two consulting firms to study and report on the potential "design, projected construction costs and necessary facts on economic factors for a multipurpose stadium to serve Seattle."[15] Each firm was lauded for its work on other stadiums across the country.

The first firm was Western Management Consultants, which provided an economic and site-selection study for potential locations. The second was Praeger, Kavanagh & Waterbury, the same firm that had just consulted on the Houston Astrodome. The county asked the Praeger firm to prepare a reliable budget cost estimate for an all-purpose stadium, hoping to resolve one of the largest misgivings about the earlier stadium proposals. While primarily for budget purposes, its report showed a schematic design of a possible stadium configuration. In March 1966, the Praeger firm submitted its report, which "made use of previous designs [that its] office developed on other major modern stadiums for baseball, football, and a combination of these sports."[16]

The report planned for a stadium with a seating capacity of forty-four thousand for baseball and forty-nine thousand for football, on both urban and suburban sites. The Praeger firm explored both an open and a permanently roofed stadium—with a structural steel dome with a ring girder along its lower edge. The Praeger report stated that a stadium with a structural steel dome on an urban site, similar to the Houston Astrodome, could be built for $32 million.[17] In the course of six years (with minimal impacts from inflation), the published cost of a stadium had gone from $15 million, to $24 million, to now $32 million. Eager to promote a modest yet realistic budget, stadium officials tempered the findings of the report. With little justification, Gandy told the *Seattle Times* that he "believed the total cost could be held to $30 million."[18]

Seattle–King County Stadium Report, March 1966

With the two reports in hand (from Western Management and Praeger), city and county officials immediately began pushing for a September

stadium-financing vote. On September 20, 1966, after a rigorous campaign for a multipurpose sports stadium, voters rejected a $38 million general obligation bond ordinance for stadium financing. The ordinance achieved 51.5 percent of the vote, short of the 60 percent needed to pass.

The day after the defeat, the *Seattle Times* wrote: "Certainly, the negative-balloting voters were not opposed to construction of a stadium. They obviously decided that the cost burden was being placed in the wrong place."[19] Despite the defeat of the stadium-funding proposal, a large-span stadium continued to be at the forefront of public discussion.

WOOD DOME DESIGN

Among all the discussions surrounding large-scale stadiums, Jack Christiansen was busy with his own designs. Christiansen could see how the construction of large-scale domes was becoming a topic of national interest. With multipurpose arenas being proposed in Minneapolis, Indianapolis, and New Orleans, Christiansen knew there would be a market for increasing longer-spanning structural systems. The Houston Astrodome had shown that a trussed, structural steel dome was an option, but Christiansen was certain that other materials—like thin shell concrete—could be used at this scale as well.

The 1960s was also a time of competition between different building materials providers, with each industry eager to demonstrate the properties of its material. Producers of aluminum, plastic, and engineered wood were eager to displace steel and concrete as the primary structural building materials and encouraged architects and engineers to consider new possibilities.[20] In 1967, Christiansen was contacted by the Tacoma-based architectural firm Harris & Reed, about a conceptual design for a large-scale dome—built entirely out of wood. James M. Harris (1928–), a graduate of the University of Oregon, and William Reed formed a partnership in 1960 and had designed several midsize projects in the south Puget Sound area. Christiansen had never worked with the firm before.

The conceptual design was sponsored by the Weyerhaeuser Corporation, a long-standing Northwest timber enterprise. Weyerhaeuser had recently initiated a design innovation series—a sequence of sponsored design challenges to architects around the country to produce new building designs that highlighted the benefits of wood construction. The design series was part of Weyerhaeuser's postwar transition from a timber company to a more products-oriented enterprise.[21] In describing the series, the company reinforced its goal: "To further explore wood's potential, Weyerhaeuser has commissioned other leading architects to develop designs ranging from commercial buildings to schools and public institutions. Among these are a child guidance center, a religious center for a university, a restaurant and cocktail lounge, and a community art center."[22]

As the tenth design in the series, Harris & Reed with Christiansen were asked to capitalize on the national interest in large, multipurpose facilities. The *Seattle Times* reported that "with the country's metropolitan areas discussing the possibility of building a domed stadium, Weyerhaeuser told the Tacoma architectural firm of Harris and Reed and the engineering firm of Worthington, Skilling, Helle and Jackson to 'show them how it can be done economically with wood.'"[23]

Weyerhaeuser Wood Dome, not built. Model view, 1967.

Harris & Reed set the architectural program for the design as a multipurpose building, fitting a baseball configuration (seating 52,000 people) and a football configuration (seating 56,500 people) under a single roof. While suggesting that the building could be located at an Interbay site (between the Queen Anne and Magnolia hills), the architects did not design for a specific site. Perfectly circular in plan, a corrugated wood dome would rest on top of inclined seating, with circulation and support services underneath. Christiansen described the first steps in the design of the wood dome:

> When it became clear that spans of as much as 800' would be required to satisfy the functional requirements of a covered stadium, we realized we were dealing with a roof structure of a scale that far exceeded that of any other known wood construction. It was evident that, in a structure of this magnitude, it was necessary to design all elements of the roof to function together structurally for maximum stiffness and strength. . . . We selected the dome as the basic structural shape because of its inherent structural efficiency and because the circular shape served the functional needs of a sports stadium quite well.[24]

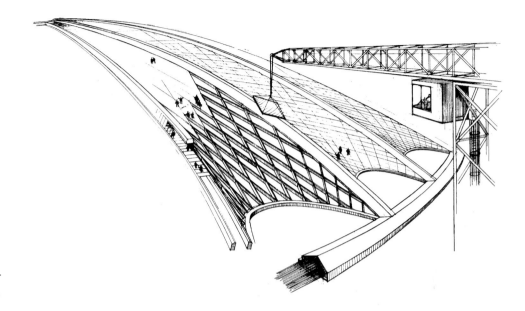

Weyerhaeuser Wood Dome, not built. Diagram of construction process, 1967.

At such a scale, new types of material and structural innovation were necessary. Christiansen began to design the structural system as several built-up elements of wood, all joined together in a monolithic dome: "We soon determined that a simple dome, by itself, was not stiff enough. Our solution was to corrugate the roof to achieve a double curvature, thus adding greatly to the stiffness and strength. The 800' diameter, circular roof is divided into 20 pie-shaped structural units. Each unit has a maximum width of 85' at the spring line. The rise of the dome is 120', providing a reasonable rise-to-span ratio of approximately 1:7."[25]

Even though he was designing a dome on an unthinkable scale, Christiansen returned to his previous design approach of dividing the vaulted surface into a series of radial segments—a similar strategy used for the Chrysler Styling Center dome. Each one of these twenty segments (with double curvature from hyperbolic paraboloid geometry) could become rigid elements individually and then be fastened together through ribs.

Christiansen described the dome as composed of four essential structural components. First, a lattice work of glulam beams, spaced 10 feet on center in two diagonal directions, would produce the hyperbolic paraboloid geometry (similar to the formwork of Christiansen's earlier thin shells). Next, a series of built-up plywood sheets, 2 inches thick, would be nailed to the glulam lattice. Weyerhaeuser publications were eager to point out that these would be "Weyerhaeuser's new prefinished aluminum overlaid panels," providing a weatherproof envelope.[26] This composite structure resulted in a "continuous thin shell–type structure, which develops great strength and stiffness because of its doubly curved shape." The geometry of this shell surface would fill the space between upturned, radial stiffening ribs. These ribs would be massive, hollow plywood box-beam sections, triangular in shape, running from the crown of the dome to the perimeter. Projecting 9 feet high above the roof, the open section of the ribs allowed for a 7-foot head height clearance inside the ribs for servicing and maintenance of the dome (e.g., fixing light fixtures and other mechanical/electrical equipment). Last, these shell segments and ribs connected at their base to glulam arches (in a horizontal plane) at the dome's spring line and a perimeter tension ring. As with all his other projects, Christiansen detailed an efficient construction sequence. He described how two opposing pie-shaped segments could be constructed simultaneously and supported on tubular metal scaffolding. Upon completion, most of the scaffolding could be lowered and moved circumferentially to the adjacent bay in order to repeat the operation. The extreme scale did not stop Christiansen from thinking about how the dome could be built.

Christiansen embraced the challenge of the design: "It has been a pleasure to participate in the development of this most interesting concept. Harris & Reed are to be commended for the design excellence of the total stadium scheme presented—it is, I believe, a thing of beauty."[27]

The wood dome attracted national attention and was immediately seen as a wood rival to the steel-framed Astrodome. In March 1967, the *New York Times* wrote that "the wooden-domed stadium will be large enough to encase the Houston Astrodome and leave room to spare. The Texas arena measures 642 feet in diameter and is 208 feet high. The Weyerhaeuser Company said its wooden-domed sports palace would measure 840 feet in diameter and be 250 feet high."[28] The article claimed that several major cities, including Philadelphia, Boston, and Minneapolis, were potentially interested in using the dome design, though none would ever be built.

The reception of the wood dome in Seattle was quickly linked to the ongoing discussion of a multipurpose stadium. In April, the *Seattle Times* quoted Lowry Wyatt, vice president in charge of wood products, as saying, "It is not the intent of Weyerhaeuser to build it—only demonstrate the architectural possibilities of wood products. . . . We are not in the business of retailing stadiums. We don't have any stock."[29] However, Joseph Gandy was quoted in the same article, stating, "We fully expect to get such a facility built in the Seattle area."

While the King County Stadium project would ultimately not be built out of wood, Christiansen had shown that it might be possible. In the years to come, however, the Northwest would get a large-scale wood dome with the construction of the Tacoma Dome in Tacoma, in 1983.[30]

Still, in the conceptual design of the Weyerhaeuser dome, Christiansen had expanded the core design ideas, started with the Chrysler Styling Center, to a massive scale. This basic concept for the roof would later be executed in concrete for the Seattle–King County Stadium. The wedged segments, combined shell-and-rib structure, and sequential construction process would all become key parts of the design.

FORWARD THRUST IN SEATTLE

By late 1967, political momentum was increasing once again for a stadium proposal in Seattle. This time, the stadium bond ordinance was included as part of an even larger, region-defining funding package: Forward Thrust. Led by the civic activist James Ellis, Forward Thrust was part of a "new cycle of urban reform," intended to transform the economic prosperity of the region into urban infrastructure.[31] The initiative included a proposed mass transit system, more public parks, and upgrading of public utilities as an investment in the long-term future and growth of the Puget Sound region.

Ellis called for cultural facilities as a key part of the Forward Thrust package and modified the stadium description from earlier proposals in this direction. While the emphasis still remained on attracting major-league sports teams, the Forward Thrust proposal recommended a true multipurpose stadium for a variety of athletic, recreation, entertainment, and convention activities. The package stated that there should be "maximum investigation of all possibilities for utilization of each part of the facility," making the stadium a cultural as well as a sporting resource.[32] By broadening its appeal, the stadium project finally found enough support within the community to proceed.

SEATTLE CENTER STADIUM

On February 13, 1968, King County citizens approved $40 million for the multipurpose stadium project, with 62 percent of the vote. A budget of $30 million was allocated to direct construction costs, with $10 million set aside for land acquisition, fees, taxes, and contingencies for the construction of the multipurpose stadium. These cost estimates were slight reductions from the recommendations of the earlier Praeger report and, following Gandy's suggestion, indicative of the interest in keeping the costs for the stadium as low as possible. Seattle finally had funding for the multipurpose stadium, even though the site was still uncertain.

Western Management Consultants' report had evaluated several sites, five in detail: Seattle Center, Yesler Way, Riverton, South Park, and Northrup Way. In light of the $40 million restriction, the report recommended the South Park site to keep the project on budget. Yet this suburban site, far from downtown Seattle, was soundly rejected by the Seattle business community. An organization of downtown Seattle businesses, called the Central Association of Seattle, wrote its own report on potential stadium sites and concluded that the Seattle Center was the best fit.[33]

The Washington State Stadium Commission, forced to choose between its own consultant's report and Seattle businesses' interests, recommended the Seattle Center as the site for the multipurpose stadium. The stadium was to be located on a triangular parcel, east of Fifth Avenue North, between Broad Street and Mercer Street. But the commission also described nine conditions that had to be satisfied before the stadium could move forward, including construction of additional parking facilities, acquisition of city-owned land, and a redesign of the proposed Bay Freeway—a high-speed link connecting Interstate 5 to the Seattle Center and Elliott Bay. The site selection was hotly debated, and demands for transparency in the decision-making process spurred public lawsuits.

Despite serious concerns, and continued uncertainty over the site, in November 1968, the county released a request for qualifications from architectural teams interested in designing the stadium. The design prospectus for the stadium explained: "It is anticipated that the stadium will be an all-weather covered facility providing space for major league baseball, major league football, and other sporting events. It will also be equipped with 'quick change features' enabling the facility to be used for conventions and entertainment purposes requiring large seating capacities and for conversion to other uses, such as large exhibitions, shows, circuses, etc."[34]

The call resulted in a large response from across the nation. Overall, twenty-one teams of architects and engineers, representing some of the most prominent firms in the United States, submitted qualifications. These included Edward D. Stone (with Stickle International as architects, engineers, and planners); Charles Luckman Associates (with Severud Associates as structural engineers); George H. Dahl (architects and engineers); Decker, Kolb & Stansfield (with Hostmark & Powell, structural engineers); Welton

Becket & Associates (architects and engineers); Durham, Anderson & Freed with Paul Thiry (with Sverdrup & Parcel and Associates, engineers); Fred Bassetti (with Finch-Heery, engineers, and Daniel Kiley, urban planner); Bindon, Wright & Partners (with Studio Nervi, architect–structural engineer); Skidmore, Owings & Merrill (Portland office, architects and engineers); and the John Graham Company (architects and engineers).[35]

This group also included the joint venture of a firm named Naramore, Skilling & Praeger. This team was the combination of three firms—Naramore, Bain, Brady & Johanson (as architects); Skilling, Helle, Christiansen & Robertson (as structural engineers); and Praeger, Kavanagh & Waterbury (as engineering consultants). In this arrangement, John Skilling, as the entrepreneurial, business-minded partner of the firm, had aligned with his friend Perry Johanson at NBBJ, as he had done many times before. In a strategic move, the two had reached an agreement with Emil Praeger—the consulting engineer who had both consulted directly on the Houston Astrodome and produced the 1966 King County–sponsored report on stadium costs. Praeger's presence on the team was important in giving the two local firms a national credibility, even though he would contribute very little to the actual stadium design.[36] Both NBBJ and the Skilling firm had designed projects all over the United States, but never a large-scale stadium, and Praeger's presence (in name only) assured the county that they could handle the large-scale project.

On February 6, 1969, four of the twenty-one teams were selected for interviews: Bassetti; Skidmore, Owings & Merrill; Thiry and Sverdrup & Parcel; and Naramore, Skilling & Praeger.[37] On March 3, the Board of King County Commissioners, acting on the recommendation of the Washington State Design Commission, selected the joint venture of Naramore, Skilling & Praeger to design a stadium for the Seattle Center site.[38] With his strategic partnership with Praeger and Johanson, Skilling had secured one of their largest design projects to date for Christiansen. His growing expertise in long-span structures would play an important role in the stadium project.

Though the project was still surrounded in uncertainty, the multipurpose nature of the stadium remained of paramount importance. Johanson reinforced this position, stating that "he believed the designers will be able to make the stadium a multipurpose operation. . . . The path leading toward construction of a stadium is long and cluttered with contractual, legal and technical problems to be solved."[39] By August 1969, the King County Council had recommended a public vote on the location of the stadium, but the design team continued with its work.[40]

In nine months, the design team produced a schematic design for Phase I of the King County Multipurpose Stadium, dated December 4, 1969. The documents were signed by Johanson, representing the joint venture of NBBJ; Skilling, Helle, Christiansen & Robertson; and Praeger, Kavanagh & Waterbury. While Johanson served as the partner-in-charge for NBBJ, the project architect was Dean Hardy, the project programmer was Charles Tinder, and the project designer was Bob Sowder (1928–). In the years to come, Sowder would emerge as the most significant design representative of the firm and work the most closely with Christiansen.

The architectural approach to the stadium was driven by function, striving toward a "clarity of expression" in the overall building form. A stated goal of the design was "to ensure that various functional requirements are organized in such a way as to produce a total project form which is a clear definition of the purpose of the building."[41]

The document describes an architectural process of analysis, synthesis, preliminary conclusions, testing, and final solution. An exhaustive "juxtaposition" of the several program demands of the multipurpose space (e.g., a baseball diamond overlaid with a football field in various orientations) began to produce a building perimeter, which defined the overall building form. Next, the team evaluated different exterior geometries that could accommodate the various perimeters—circle, oval, squares of different sizes. During this synthesis, the design team concluded that a parallelogram or square building would best suit the multipurpose nature and the current stadium site. A circle, the first geometry explored, was discarded because it could not be related to the size and shape of the stadium site.

The team then presented a graphical review of the recent professional sports stadiums built across the country, showing their geometries and seating configurations. This study included the existing Sick's Stadium (the baseball stadium in Seattle), the Houston Astrodome, Dodger Stadium, the Memorial Coliseum in Los Angeles, the University of Washington Husky Stadium, and the "Louisiana Stadium" in New Orleans.

After describing the study, the design team presented its final solution. Prioritizing sight lines from the seats to the playing field, and maximum flexibility, the design team settled on a plan that was roughly 650 feet by 650 feet, with the corners cropped at 45-degree angles, giving the building the appearance of a symmetrical, unequal-sided octagon. Only after this architectural study was completed were the structural aspects of the stadium discussed, revealing Christiansen's active involvement on the project. The

Seattle Center Stadium Design, not built. Sketch of stadium roof with Mount Rainier and Space Needle, 1969.

design document read: "The octagonal plan suggests a segmental system consisting of a number of triangular shaped and/or diamond shaped panels, with major ribs or stiffeners along the lines of panel intersection."[42]

The document presented three different structural options for constructing the long-span roof—steel, wood, and concrete. The steel option would have used a series of flat and triangular trusses to cover the long-span space, with metal deck roofing on top. A simple design, the trusses defined large, flat planes on the roof surface—a "truncated, octagonal pyramid." Given the building's plan, a rounded dome was not possible, meaning that the structural solution would be distinctly different from the steel roof of the Houston Astrodome and the option presented in Praeger's original stadium report. The straightforward description in the Phase I document does not describe the advantages of a steel-framed roof, highlighting only the additional need for fireproofing on the outermost steelwork. Still, the steel option was shown in all the renderings of the buildings in the Phase I document.

The concrete option allowed Christiansen's experience to shine through. The description read: "This segmental, vaulted thin shell structure consists of two types of doubly curved, thin shell panels, triangular panels and diamond shaped panels. These panels join together at ridges and valleys where stiffeners are provided. The valley stiffeners gather the thrusts from the vaulted panels and deliver them to the perimeter columns. The horizontal thrusts are resisted by prestressed cable ties between the tops of the major piers and by a continuous perimeter tie."[43]

In contrast to the dispassionate description of the steel option, the thin shell concrete description becomes more convincing: "This scheme presents

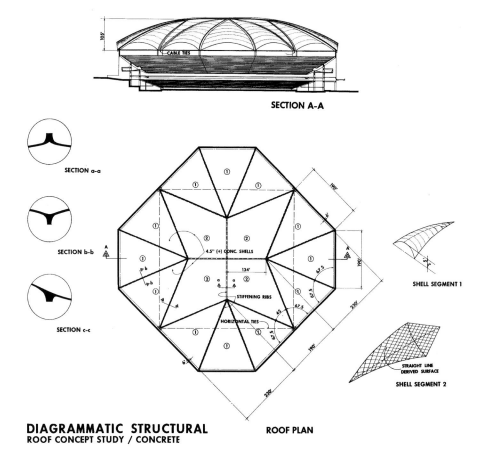

SECTION A-A

SECTION a-a

SECTION b-b

SECTION c-c

4.5" (+) CONC. SHELLS

134'

STIFFENING RIBS

HORIZONTAL TIES

SHELL SEGMENT 1

STRAIGHT LINE
DERIVED SURFACE

SHELL SEGMENT 2

DIAGRAMMATIC STRUCTURAL
ROOF CONCEPT STUDY / CONCRETE

ROOF PLAN

Seattle Center Stadium Design, not built. Christiansen's design for thin shell concrete roof with stiffening ribs, 1969.

a smooth clean structure that is fireproof and essentially maintenance free throughout. The repetition of but two different panels allows for a maximum reuse of formwork with consequent cost savings."[44]

Though attempting to objectively present the different options, the preference for concrete is evident. Christiansen's previous work combining warped segments of shells was put to good use. With no control over the footprint of the building, Christiansen was able to subdivide the geometry into segments and develop a structurally stable, buildable roof out of only 4.5 inches of concrete. The curving geometry of the roof created a striking profile as well. The roof covered the multipurpose functions inscribed in the lower square, yet appeared as a rounded—almost dome-like—form.

The design team also explored a structural solution in wood—an option that never would have been considered if not for Christiansen's earlier design with Harris & Reed. The wood scheme provided a similar layout of ridges and valleys as the concrete option, which was significantly different from the circular dome Christiansen designed before, yet the structural materials were nearly the same.

The wood design called for an identical use of a "two-way grid or lamella of glulaminated beams" sandwiched in between prefabricated plywood panels. The primary difference was the stiffening ribs—similar upturned, triangular hollow boxes—made of steel trusses, not wood. This was likely required because of the unique configuration of panels, inducing higher moments in each member. Also, to meet fireproofing standards, the design required a continuous prestressed concrete perimeter slab.

Phase I of the schematic design was approved, and the joint venture team moved ahead with design. Yet the placement of the stadium at the Seattle Center site was becoming more and more tenuous. In December 1970 and January 1971, the county held a series of public presentations of the drawings for the stadium and the proposed Bay Freeway.[45] Still, public support for the Seattle Center site continued to erode as the design neared completion.

Determined to move ahead, the county initiated Phase II of the design on February 16, 1970, authorizing Naramore, Skilling & Praeger to continue to develop its designs for the Seattle Center site. With this schedule, construction could begin by the end of the year and produce a stadium ready for the 1973 baseball season. County executives warned that any further delay would jeopardize completing the project on time and, most importantly, within the $40 million authorized by voters.[46]

Despite these pressing concerns, between February and March 1970, King County executive John Spellman was overwhelmed with requests from the public to stop the stadium and Bay Freeway project. Department of Transportation officials lobbied the county to stop the project, claiming that the traffic congestion in the area would be too great. With a vote to eliminate the Seattle Center as a stadium site on the ballot in April 1970, the county began to consider suspension of the stadium design project.[47]

Just as the site of the stadium was becoming less certain, the decision to use thin shell concrete was finalized. On Tuesday, April 7, 1970, at a meeting of the Stadium Steering Committee, Johanson reported that the design team had recommended, after very careful review, the use of a concrete roof for the stadium dome. He pointed out that the aesthetics and the cost (approximately $1 million less than other materials) dictated this decision.[48] For Christiansen, the stadium was both a validation of his previous work and a new frontier for his thin shell concrete designs.

With the new roof profile set, on Friday, May 15, 1970, the King County Design Commission gave unanimous approval to the final design for the "domed stadium at the Seattle Center."[49] The *Seattle Times* reported:

A new shape for the roof of the proposed King County domed stadium was revealed today as representatives of Naramore-Skilling-Praeger presented their final design-development documents to the county's design commission. The roof proposed for construction is a series of segmental concrete arches with a thin shell of concrete suspended between them. In earlier drawings, before concrete was chosen for the dome, the roof was shown as a series of pitched panels sloping from side walls to a flat area at the top. Architects said the thin-shell concrete roof, which would be 9½ acres in size, is the most economical way to cover the stadium.[50]

The design was revealed through a series of large renderings published in the *Seattle Times*, perhaps intended to sway public opinion toward approval. Several civic leaders including Joseph Gandy and architect Perry Johanson, hoping to rally excitement for the stadium, endorsed the site for the domed stadium at the Seattle Center.[51]

Seattle Center Stadium Design, not built. Sketch of potential interior condition, 1969.

But the new design publicity and public campaigning were not enough. Just four days later, on May 19, 1970, the ballot initiative "Ordinance Prohibiting the Location of the Multipurpose Stadium at the Seattle Center" passed. This "yes" vote meant that the domed stadium had to be relocated and the design proposal scrapped. The design team of Naramore, Skilling & Praeger, having fulfilled its schematic design tasks, was not authorized to proceed, effectively removing the team from the stadium project all together.

KING STREET STADIUM

The vote rejecting the Seattle Center site forced a wholesale reconsideration of the entire stadium project, at a time when the economy of the Seattle region was in rapid decline. In the midst of another "Boeing bust," Seattle was deeply affected by the aerospace giant's reduced government contracts and by increased competition from Lockheed Martin and McDonnell-Douglas. The number of employees at Boeing had declined rapidly from a peak of over one hundred thousand in 1967 to less than thirty-nine thousand in 1971.[52]

In addition, construction costs nationwide were spiking. After several years of modest annual increases (3–5 percent in the early 1960s), construction costs jumped nearly 9 percent from 1968 to 1969.[53] From 1969 to 1970, construction costs increased nearly 13 percent, far outpacing inflation. With these unexpected increases, the buying power of the approved $30 million for the stadium had shrunk drastically. This loss of economic prosperity rippled across the region and heightened budget scrutiny for the stadium project.

Following the vote, the governor appointed an entirely new Washington State Stadium Commission, with James L. Wilson replacing Joseph Gandy as chairman. The commission immediately began reevaluating a multitude of possible sites, referring back to the Western Management Consultants report, while remaining open to input from downtown businesses. On January 20, 1971, the Washington State Stadium Commission voted unanimously to locate the stadium at a site south of downtown Seattle, in an existing railroad yard—known as the King Street site.

By February, the county was exploring working with other architects for the King Street site.[54] But many remained hopeful that the team of Naramore, Skilling & Praeger would be retained to "reduce costs and time in the design of another stadium."[55] Christiansen, along with Skilling and Johanson, remained active in design discussions, and the team maintained its choice of thin shell concrete as the most economical structure.

In summer 1971, surrounded by the delays and uncertainty of the stadium project and also burned out from the challenges of working in a large office, Christiansen took a ten-week trip to western Europe. Christiansen had always wanted to travel internationally and was willing to do anything to go. Not wanting to burden the firm with his absence, he informed the partners that he would be taking the trip without pay.

Traveling by Volkswagen bus, the Christiansen family made a grand tour of France, Spain, Germany, Switzerland, and Italy. The trip's destinations had a dual focus: high-altitude mountains for climbing and monuments of architectural history. Christiansen sought prominent peaks throughout Europe, climbing the Matterhorn in Switzerland (14,692 feet) and Mulhacén in Spain (11,413 feet).

He also visited many of the seminal works of Western architecture, tracking each one down with the same fervor with which he sought new peaks. In Spain, he visited the Alhambra, admiring the unique vaulting of the Hall of Two Sisters. In Italy, he saw the Leaning Tower of Pisa, St. Mark's Cathedral in Venice, and Brunelleschi's dome above the Duomo in Florence. He was particularly taken with Brunelleschi's dome, photographing its form from a distance, in detail, and from the inside. As a dome of radial segments separated by upstanding ribs, Brunelleschi's structure had a similar concept to Christiansen's earlier domes—including the Chrysler Styling Center and the Weyerhaeuser Wood Dome. He took photographs of the sectional diagrams and possible construction methods exhibited in the Duomo museum.

In France, he saw the lightness of Chartres Cathedral, the flying buttresses of Notre-Dame de Paris, and the lattice work of the Eiffel Tower. But Christiansen was also interested in contemporary work in reinforced concrete by European designers. He visited the triangular-in-plan Center of New Industries and Technologies (CNIT, designed by Bernard Zehrfuss, Jean de Mailly, Robert Camelot, Jean Prouvé, and Nicolas Esquillan and completed in 1958), located in La Defense, a neighborhood of Paris. Photographing the

Jack Christiansen underneath the Eiffel Tower, 1971

CNIT Exhibition Hall, Paris, 1971. Designed by Bernard Zehrfuss, Jean de Mailly, Robert Camelot, Jean Prouvé, and Nicolas Esquillan.

exterior, Christiansen noted (on the developed slide) the maximum span of the three-pier supported structure as 731 feet. The CNIT, also built in a self-supporting, segmental manner as Christiansen had envisioned, was one of the only long-span structures in concrete that exceeded the span of the King County Stadium, through the use of a double-shell structure.

Christiansen also visited the headquarters of the United Nations Educational, Scientific, and Cultural Organization (UNESCO), designed by Zehrfuss, Marcel Breuer, and Pier Luigi Nervi (also completed in 1958). Nervi used a concrete folded-plate structure to construct the long-span auditorium and created a library framed by tapered concrete piers. Christiansen took several photos, highlighting the expressed concrete forms.

Christiansen was clearly interested in comparing his own designs to these European works. Upon returning home, he organized the slides of his trip—with the CNIT and UNESCO side by side, followed by an image of his own Mercer Island Multipurpose Room (also completed in 1958). Both designers were older than Christiansen (Nervi by thirty-eight years, Esquillan by twenty-seven years), yet he clearly saw himself as working within the same design tradition. Through his work, Christiansen began to feel connected to other design engineers around the world.

For Christiansen, this European trip reaffirmed his design sensibilities. By seeing these globally significant works, he understood the worldwide legacy of structural engineering and began to place himself within it. He saw similarities between his own thin shell work and the dome and vaults throughout western Europe. Christiansen returned to Seattle reinvigorated and ready for the significant challenges that the King County Stadium would bring.

STADIUM RESUMES

On November 30, 1971, King County published an initiative stating that the county would enter contract with a design team, acquire land, and begin site preparation at the King Street site.[56] In early December 1971, King County exercised its option to buy the King Street site, and on December 8, King County executive John Spellman removed the first railroad spike from the site, marking the beginning of stadium work. Shortly thereafter, Spellman announced that the county had signed a new contract with Naramore, Skilling & Praeger for the design of the county's multipurpose stadium—roughly a year and a half after the Seattle Center rejection.

Though the same design team was retained, the King County Stadium project was now fundamentally different from before. The original design of the stadium at the Seattle Center site would have been carried out through a series of orderly design phases, with ample time for development and feedback. In the face of inflating costs, the design of the stadium at the King Street site, however, would now be accelerated, with the aim of getting a stadium built as soon as possible.

The design team also discovered that the Seattle Center design could not simply be relocated to the King Street site, for a variety of reasons. Most importantly, the stadium now had to meet different programmatic demands. As part of its multipurpose mandate, the stadium would still need to host conventions and exhibitions, but now was unable to use the existing parking and ancillary event space at the Seattle Center. This meant that the stadium needed to provide more square footage.[57] Moreover, the NFL had increased the number of seats required to host a professional football team, from fifty-five thousand to sixty-five thousand.[58]

In addition to programmatic changes, the budget for the project continued to be affected. Construction costs rose another 13 percent between 1970 and 1971, far outpacing inflation of the bond value (which remained at roughly 3.5 percent) and further diminishing the value of the committed funds. With a heightened sense of public scrutiny after the unsuccessful Seattle Center site, Spellman reassured the public that the construction budget remained the same: $30 million in 1968 dollars.[59] Spellman was also actively engaged in the pursuit of an NFL franchise, and the relocation from Seattle Center had added at least two years to the timeline for a football-ready stadium. Spellman pressed the design team to have the stadium open for the 1974 football season.

Thus, on this new King Street site, time and cost were the driving factors of the King County Stadium project. With this focus, the design of the stadium shifted. Success on the project was redefined to be quantifiable metrics of dollars spent and time of construction. Despite these challenges, the design team remained dedicated to the task at hand, stating that "preliminary cost studies indicated that by careful control of the design and the elimination of all unnecessary frills, it might be possible to do the facility within budget."[60]

The King Street stadium would be a stripped-down version of the earlier stadium proposal. The extreme cost-consciousness would affect every decision made surrounding moving the stadium forward—from planning, to design, to construction. One concern was the contractual agreement between the county and the general contractor: "Some skepticism existed on the part of the County Council based upon knowledge that all such stadium facilities that had been either constructed or planned had costs running well above $30 million. In order to ease the minds of the County Council and Design Commission, the Design Team suggested the scheme of pre-bidding a number of large, significant cost items prior to the total completion of the plans, so that a very close check could be kept on costs."[61]

This construction arrangement was referred to as a "fast track" contract. The agreement meant that as the design team finalized the different parts of the stadium (foundation plan, column sizes, etc.) these parts of the project would be sent out for bid from interested contractors. In this manner, the stadium could be divided up and built by any number of builders bidding the lowest price for a particular piece. The county saw the contract structure as a way to speed up the project, since contractors could begin planning for the construction of various parts before the rest of the building was completed. It would also allow the county to monitor the most expensive components of the building and ensure that the project stayed on budget.

In its contract with the county, Naramore, Skilling & Praeger was required to complete the design of the stadium (drawings and specifications for general contract bidding) in six months. That was about half the time usually required for a facility as large and complex as the stadium. Yet the team began providing bid package documents even earlier than that.

The speed with which the design progressed led to simplifications of the Seattle Center design. One of the first items designed for the stadium was a circular dome. On December 29, less than three weeks after Naramore, Skilling

& Praeger re-signed, the *Seattle Times* reported that the county stadium would have "a circular design and probably a thin concrete dome."[62] With more space surrounding the stadium than at the Seattle Center site, the design team likely revived the "circle" configuration plan from the Phase I report.

A circular plan enabled a more typical dome-shaped profile for the roof. With a circular perimeter, the outward thrusts from the roof could be easily handled by a continuous tension ring at the base of the dome—minimizing the moments generated and eliminating the need for external buttressing. In short, the circular plan was the cheapest option.

Other programmatic demands supported a dome configuration as well. The additional interior height that the dome created—the "rise" of the dome—could be used to satisfy the minimum 200-foot clearance height required by Major League Baseball.[63] Also in the area between the circumference of the dome and the differing playing fields within (baseball diamond, football field, etc.), the designers could place an even distribution of seating. With cost at the forefront, the *Seattle Times* reported: "A thin con-

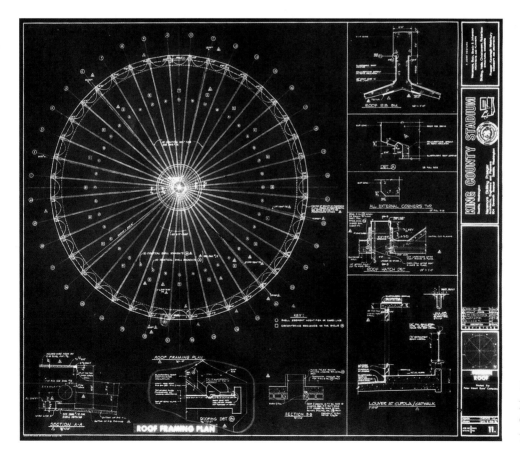

King County Multipurpose Stadium, Seattle. Concrete roof plan, 1972.

crete dome, with an average thickness of six inches, shows promise of being the most economical, John V. Christiansen, design-team representative said. Christiansen added, however, that bids would probably be received for wood and steel domes."[64]

Once again, Christiansen had taken the design lead of the long-span structure and kept thin shell concrete at the forefront of the discussion. Again, as in the Seattle Center design, Christiansen was open to other structural material choices but stayed committed to a thin shell structure.

Just over a month after signing the contract, the design team finalized the stadium design. On January 21, 1972, the King County Design Commission approved the stadium design documents for a "domed stadium."[65] Perry Johanson stated that "all major design problems for the building appear to have been resolved," and Bob Sowder affirmed that "the project is exactly on schedule."[66]

In fact, the roof design was the first bid package to be completed. The choice of roof system was essential in setting the design of the entire stadium

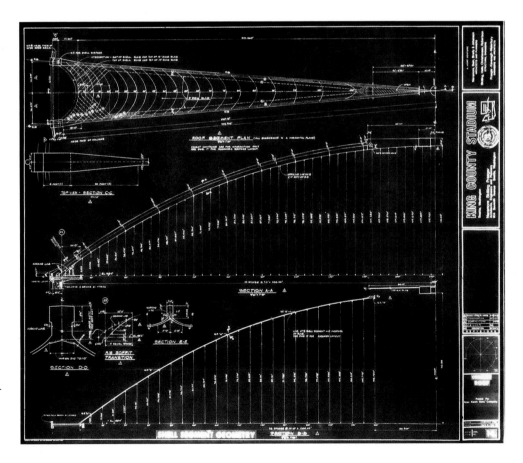

King County Multipurpose Stadium, Seattle. Concrete roof details, 1972.

below, and the design team would accept proposals in any material. The *Seattle Times* reported: "The roof system is being handled first so that it can be incorporated into the design of the entire stadium."[67] Depending on the lowest bid for the roof construction, the rest of the stadium would adjust its design to suit the selected structural system.

On February 28, the county held a pre-bid conference for the stadium roof. Christiansen had prepared the bid documents in less than three months' time. While the team remained officially open to the possibility of alternative materials, the drawings were labeled "Concrete Roof." Christiansen provided a detailed design for a concrete option, complete with structural sizing, materials, and a possible construction method. He also provided the functional requirements that the dome needed to satisfy (span, loading, etc.), leaving the dome design in other materials (e.g., steel, wood) up to the bidding contactors.

Christiansen's thin shell concrete design related more directly to the Weyerhaeuser Wood Dome and the Chrysler Styling Center than to the Phase I stadium at the Seattle Center site. Working from a circular plan, Christiansen divided the roof into forty wedge-shaped segments. Between each segment, large, 6-foot-tall, upturned concrete arch beams acted as stiffening ribs. As before, each thin shell segment was defined by hyperbolic paraboloid geometry, with straight-line formwork shaping the double curvature between each rib and requiring only 6 inches of concrete thickness. The dome had a clear span of 661 feet 4 inches and a rise of 110 feet. The shell and ribs delivered thrusts to a 24-foot-wide tension ring at the base of the dome. The entire dome rested on top of 140-foot-high roof support columns.[68]

King County Multipurpose Stadium, Seattle. Architectural model, 1972.

Structurally, the ribs and shell would work together to resist gravity, wind, and earthquake loads. The ribs would work as simple arches to carry the self-weight of the roof, braced by the shell on either side. The shell would also provide continuity between ribs, engaging membrane behavior as a dome to resist live loads.[69] This behavior also made it possible to cast the dome segmentally—with pie-shaped wedges supporting their own weight until the entire dome was cast.

Christiansen had defined sixteen segments on the Chrysler dome and twenty segments on the Weyerhaeuser dome—using wider segments to complete the circular dome. This increase in the number of segments would make each wedge smaller and allow for even more repetitive use of formwork. The contractor for the Chrysler dome had found it more economical to cast the entire dome at once, but Christiansen made sure that would not be the case on the King County Stadium.

In the bid documents, Christiansen detailed several procedures for pouring the concrete roof—easily the most expensive component of the design. He provided an option for filling the entire domed space with scaffolding and pouring the dome all at once. But on his suggested shoring layout, Christiansen designed a system of movable forms that took advantage of the repetitive geometry of the roof. A central support tower constructed of steel was built exactly in the center of the dome. Four smaller towers—at the quarter point of the circle—were built roughly 200 feet out from the center, on top of rails, allowing them to rotate around the center tower. A long-span truss ran over the top of these towers, completing the span from the center to the exterior perimeter and supporting the wet concrete. Construction sequencing played a large role in Christiansen's design: "The King County Stadium dome was conceived (in part) as a stiffened, segmental dome because the segmental construction allowed for step-by-step construction with potential re-use of forms and consequent cost-savings."[70]

With forty identical segments, the dome could be built with only four support towers (plus the central support) and four sets of forms. By comparison, the Houston Astrodome required thirty-seven support towers to hold up the steel framework during construction. With a tight cycle of forming, pouring, curing, and moving, the dome could be built in a short amount of time. With a tightly sequenced construction, the entire dome could be built in only ten pour cycles. Christiansen knew that if a contractor used the efficient forming system, the concrete dome—with its combination of material efficiency and systematic construction—could be built for a low price. The only thing left to do was to find the right contractor.

On April 12, 1972, the county evaluated bids for the stadium roof, including from the Peter Kiewit Sons Company to build a precast concrete dome of its own design for $5.8 million and Christiansen's dome for $6.6 million. The Donald M. Drake Company of Portland bid $5.9 million to build the 660-diameter concrete dome. Despite the $100,000 difference, on May 12, Drake's bid was accepted.[71] The precast concrete dome that Kiewit proposed was "not acceptable" because it depended on the county entering into a separate contract with Kiewit's design consultant (unnamed). Other, lighter dome solutions were considered, but the county decided that fireproofing requirements, insurance, and increased maintenance costs would have offset their savings in weight.

The Donald M. Drake Company had previously worked on roadwork, grading, and transportation projects and had only limited experience on buildings. In 1971, it built the Washington State Highway Administration Building, but its primary expertise was in building bridges and civic infrastructure. Most significantly, Drake had previously worked with Christiansen on the construction of the Nalley Valley Viaduct in Tacoma. Even with the King County Stadium's scale exceeding many of Drake's transportation infrastructure projects, the company hoped its experience forming long-span concrete bridges would translate to long-span concrete roofs.

Once the use of Christiansen's concrete dome had been finalized with the county's acceptance of Drake's bid on May 12, 1972, the Naramore, Skilling & Praeger design team could complete the rest of the stadium design. The design team worked twelve-hour days. Just before the drawings were completed, the number of architects and engineers on the project reached 135. Less than two months later, the final design drawings were ready for bid, dated June 30, 1972. These included complete architectural, structural, mechanical, and electrical drawing sets, ready for the next phase of bidding.[72]

The building was designed to be massive, covering more than 9 acres of area and enclosing over 67 million cubic feet of volume.[73] In their search for ultimate economy, the architects challenged Christiansen to minimize the number of structural elements needed below the concrete dome as well. They then set to work making each part of the structure perform multiple functional roles, again, to reduce costs.

For example, the forty columns that supported the concrete dome (one under each dome rib) were 6 feet wide and 8 feet deep but were cast into a U shape with a 3-foot-wide, open channel running vertically up the building. This channel, easily closed off with a metal panel, served as a vertical chase—and concealed electrical conduits, roof drains, and other building systems running up and down the building. Additional panels within the channel even allowed these channels to become air plenums—the conduit for pumping fresh air throughout the building.

Also, the exterior pedestrian ramps that circled the building provided "the simplest, quickest and least expensive way to get 65,000 spectators where they want to go."[74] These ramps also braced the perimeter columns against buckling and were an essential component in providing lateral stability against wind and earthquake forces.

The extreme economy with which every component was designed kept the building costs in line. The mechanical system, concession stands, scoreboard, seating arrangement, and other building elements were modified or eliminated to save cost.[75] The "fast track" process allowed each component to be bid separately. Christiansen recalled: "Using this process, and prior to the completion of all the design documents, a contract for a dome at 5.92 million dollars, a contract for 34 giant mechanical air-handling component units at 1.57 million, a contract for all the precast work at 2.23 million dollars, and a contract for $670,000 for foundation piling were secured. The sum of these pre-bid items was approximately 35% of the total estimated cost; and since all were within the estimate, the County elected to continue with the project."[76]

The primary design architect, Bob Sowder, described the architectural expression in the design: "We've tried to express this stadium structure in its boldest form. It's like a bridge where the function is dramatically apparent."[77] Sowder recognized how the expediency and cost savings had resulted in an unapologetic, exposed structural form. In reality, the design simply put Christiansen's structural design on display on a massive scale, heavily dictated by the severe financial constraints of the project. Though he officially remained a consulting engineer on the stadium project, Christiansen was considered one of its primary designers.

A publication from King County promoting the stadium made Sowder and Christiansen the face of the stadium design: "Bob Sowder and Jack Christiansen dressed in business suits, sitting in offices full of blueprints, don't look like gridiron heroes. But they quarterbacked the design team for the King County Stadium to a decisive win against the tough opposing teams of Time and Cost."[78] The publication went on to describe each designer:

> Bob is a solid bundle of energy whose motor always seems to be running at full speed. His bulldog tenacity is counterbalanced by a ready laugh and contagious enthusiasm.
>
> Jack is tall and lanky; friendly but sparse of word. He looks and acts a little like a laconic cowboy, unimpressed by his considerable skill, modest about recently being named Engineer of the Year by the Consulting Engineers Council of Washington, and collecting a national honor award from the Consulting Engineers Council.

On August 5, 1972, the Donald M. Drake Company submitted another bid to build the entire stadium (in addition to the concrete roof) for $29.3 million.[79] The extreme cost-cutting measures had paid off, and the design had been bid within 0.2 percent of the expected cost projections. This was a remarkable feat, especially when compared to the cost of other multipurpose facilities. The Houston Astrodome had ten thousand fewer seats and was built for $35 million in 1964, which equated to roughly $67 million in 1972.[80] Other stadiums of similar size being developed at that time, including the Montreal Olympic Stadium, were in the $50 million to $80 million range. The Superdome in New Orleans was originally estimated to cost $35 million, but its budget had drastically expanded to over $80 million and was increasing.

With a contractor on board, on October 16, 1972, the City of Seattle issued a building permit for the stadium, and on November 2, King County executive John Spellman led the ground-breaking ceremony.

CONTINUED ROOF ANALYSIS

Even though the concrete roof was the first structural component to be bid out, it would be the last structural component built—a time difference that Christiansen would use to his advantage. In less than three months (from the time he restarted the stadium project to the bid document date), Christiansen had specified the largest, freestanding concrete dome in the world, yet questions still remained. The dome's more than 600-foot span dwarfed that of the Chrysler Styling Center and was over twice as long as the Rivergate span. Only the Weyerhaeuser dome, at 800 feet in diameter, proposed a larger span—and that project never left the conceptual stage. Despite the fact that the bid documents had already been sent out, Christiansen continued to study and analyze the dome structure in the office.

In the months that followed, while other parts of the project were moving forward, Christiansen conducted a wide range of analyses for the concrete roof. As with Rivergate, Christiansen used several different methods to investigate the structural behavior of the dome, including both computer models and physical models.

COMPUTER ANALYSIS

Both the Seattle and New York offices of Skilling, Helle, Christiansen & Robertson had continued to expand their own in-house computer capability. Christiansen continued to advance computer analysis techniques, both drawing on industry advancements and writing his own programs. In the late 1960s, Christiansen began to use a state-of-the-art analysis program called STRUDL (STRUctural Design Language), developed by a group at MIT.[81] This group produced programs such as ROADS (ROadway Analysis and Design System) and STRESS (STRuctural Engineering System Solver) and, in 1967, released STRUDL.[82] STRUDL, developed by John Biggs and Robert Logcher, was an extension of the earlier STRESS program as a frame-analysis system, but it now allowed individual design engineers to modify the input and output data to suit specific, structural engineering project definitions.[83]

Christiansen had been a central part of integrating these computer analyses into the design process at the firm. By 1971, only four years after the release of STRUDL, Christiansen, Leslie E. Robertson, and Richard E. Taylor programmed their own modifications to suit the design of buildings in the Skilling, Helle, Christiansen & Robertson office.[84] Working together, the group created a supplemental program to the STRUDL platform, which they called ADL (A Design Language).[85] This program allowed engineers to modify and redesign specific elements of a building, in response to changes in the project requirements, without losing the analysis performed previously. In a joint article, Christiansen, Robertson, and Taylor wrote: "The designer has complete control over the design process at all times. He determines the model of the building to be designed and he is free to change this model at any stage in the design process. The authors consider ADL to be a complete computer representation of the state of the art of the structural analysis and design of buildings."[86]

Also by the late 1960s, a new technique was fundamentally changing the way structural engineers analyzed buildings. The finite element method, developed by Ray W. Clough (1920–2016), John Argyris (1913–2004), and others at the University of California, Berkeley, provided a more detailed understanding of complex structures by dividing a structure (or shell surface) into smaller, discrete elements. After applying loads to these "finite elements," the computer could then reassemble them into the structural form and provide a more detailed description of stresses. The finite element method was beginning to replace the earlier frame analysis and provide a more exact assessment of structural forms.

By 1973, the office had upgraded to an IBM 1130, and Christiansen continued to write his own analysis program in the base language FORTRAN.[87] Christiansen put this advancing computer expertise to use on the assessment of the Kingdome form. Similar to his analysis of the Rivergate exhibition hall, Christiansen's segmental dome design contained repetition of identical elements and lines of symmetry, lending itself to a rapid analysis procedure. With the repetitive segments in a circular pattern, Christiansen could assume that under uniform gravity loading each segment behaved identically to every other one. In addition, each segment was symmetrical along its center line.

This geometry meant that the computer model of the entire roof could be reduced to one-half of one of the forty segments. This efficiency allowed for a much more detailed understanding of the half-shell segment and thus the overall behavior of the largest dome in the world. Embracing the most up-to-date analysis techniques, Christiansen made a finite element model of the half-shell segment, breaking the rib and shell structure into small, triangular elements and evaluating the stress state of each one under gravity loads. The firm also conducted an advanced, dynamic earthquake analysis of the building, in a separate finite element model.[88]

In addition to the hand calculations and the computer modeling, Christiansen also oversaw the testing of a physical model of one segment of the dome. The test was performed in secret, in order not to arouse suspicions in the public, and took place in the basement of a downtown building between January and May 1973. The test was voluntary on the part of Christiansen and Skilling (not required by the county), and the firm paid for the test out of its own design fees. Again, even though the design had been finalized and the bid accepted, Christiansen and others at the Skilling firm wanted to test the design before construction began to verify the computer-generated output.

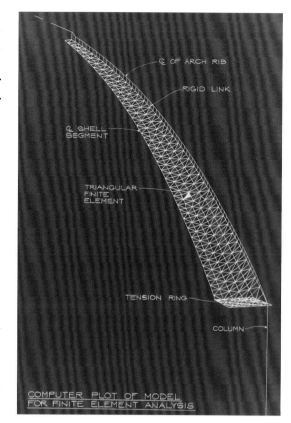

Finite element analysis segment for computer analysis of King County Multipurpose Stadium, 1973

According to Christiansen: "Although some analytical methods have been checked by model testing, the configuration of the Kingdome had not been previously tested and it was felt that it would be desirable to do so."[89] One shell segment (or wedge) was built to one-tenth the actual size. With the dome spanning 660 feet, each segment spanned 330 feet from the edge to the center point, thus the model segment spanned 33 feet. The shell segment was loaded with 25 pounds bags of steel punchings to simulate the equivalent of 160 pounds per square foot—roughly two and a half times the design loading on the stadium. The model was loaded and checked for cracking at different time intervals. "Dome Design Beats Test," read a May 1973 *Seattle Times* headline: "Five months of testing a scale model of a section of the King County Stadium's unusual dome—a design first—have been completed and to no one's surprise, it works."[90]

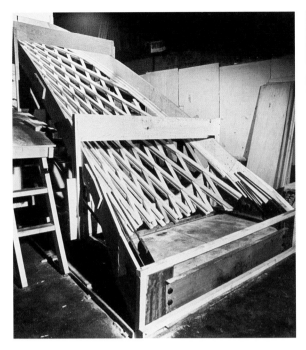

King County Multipurpose Stadium, 1973. (*left*) Formwork for physical model segment. (*above*) Physical model segment for testing structural behavior.

The concrete dome—the longest-spanning concrete dome in the world—had been verified through multiple methods, with virtually no alterations to the original design needed. This was a testament to Christiansen's original design and his years of experience with thin shell concrete construction. With the additional analysis and testing, Christiansen was more confident than ever that the dome would perform as expected.

These results were particularly welcome, as construction of the stadium was already under way. King County regularly published a newsletter, answering common questions from the public and reporting the status of the stadium construction.[91] In early 1973, the first phases of construction went smoothly, as the site was cleared and piles were driven for the foundation. In March 1973, the county announced that construction was proceeding, while also describing the courtroom victories that paved the way for construction, statistics on the project structure, and the ground-breaking ceremony. By July 1973, the county proclaimed in a headline "Stadium Project Shifts into High Gear," with the pouring of the ring of perimeter columns. By February 1974, the perimeter columns were connected by perimeter ramps, and the county declared the stadium a "new face" on the Seattle skyline.

With the stadium nearly 50 percent complete, the progress solidified commitments from professional sports teams.[92] In February 1974, Seattle

King County Multipurpose Stadium, Seattle. (*top*) Formwork segment, 1974. (*bottom*)
Scaffolding and formwork for single segment in place, 1974.

180

King County Multipurpose Stadium, Seattle. (*top left*) Four segments cast on revolving forms, 1974. (*top right*) Newsletter, November 1974. (*bottom*) Exterior view, with four segments in place, 1974.

landed its first professional sports team for the stadium—a soccer franchise, soon named the Seattle Sounders, in the North American Soccer League. And on June 4, 1974, the NFL owners voted to award Seattle an expansion franchise—creating the Seattle Seahawks.[93] Major League Baseball had awarded Seattle an expansion franchise in 1969, but that team (the Seattle Pilots) was sold and relocated to Milwaukee after just one season. The City of Seattle, King County, and the State of Washington sued Major League Baseball for breach of contract and were finally awarded another expansion franchise in 1976—creating the Seattle Mariners.

By July 1974, the perimeter columns, ramps, and tension ring to support the dome were complete and the scaffolding and support towers to cast the dome were coming into place.[94] Excitement continued to build into November when the first four roof segments were cast, and the profile of the dome began to take shape. The King County newsletter proclaimed that the stadium would become "the West's first multipurpose, covered stadium."[95]

CONSTRUCTION WOES

The excitement surrounding the stadium, however, soon came to a halt. In December 1974, the *Seattle Times* reported: "The issue first arose more than a month ago when Drake tried to free forms from the first of four ribs it had poured for the dome. Although Drake had provided for a 2-inch drop of the falsework to free the forms from the dome, part of the 300-ton roof segment still rests on part of the form near the dome's edge. Who is responsible for solving this problem is a major element in the dispute."[96]

The issue appeared to involve the essential formwork system for casting the dome. This system required that the formwork be dropped clear of the concrete segment before rotating it to the next position. At Christiansen's suggestion, Drake had installed sand jacks to lower the formwork. These jacks essentially consisted of large steel boxes filled with sand, on top of which the shoring towers were built. When it was time to lower the forms, the sides of the boxes could be removed, allowing sand to pour out of the box and the tower to be safely lowered to the ground. When Drake had removed the sides, the company claimed, the tower had not dropped enough to clear the bottom side of the segment.

Drake insisted that the roof segment was deflecting (bending downward) more than the designers had planned on. This behavior was keeping the forms in place, and thus preventing Drake from completing the project. Skilling and Christiansen countered that the roof was performing precisely as expected, but Drake had not allowed the towers to drop sufficiently. Much of the discussion revolved around the sand jacks, including a *Seattle Times* article titled "Stadium: The House That Jacks Build."[97] Christiansen was repeatedly quoted in the local media, forced to reassert his claim that there was nothing wrong with the design of the stadium.

In reality, the work stoppage was the culmination of several events that had caused Drake to lose money. By December 1974, inflation had spiked to 12.4 percent. The OPEC embargo beginning in October 1973 had caused the price of oil to spike, from $3 per barrel to over $12 per barrel, and resulted in the reduced availability of fuel. Design changes instigated or approved by the county and strikes by workers only compounded Drake's financial position on the stadium project. Attempting to leverage the stuck formwork to its advantage, Drake wanted its contract changed to a cost-plus basis, with a minimum of $8 million in additional fees. King County executive John Spellman agreed to only $1.5 million. After talks broke down, Drake stopped work and began removing its equipment and workers from the construction site. Drake filed a $10.5 million claim against King County, and the county responded with a countersuit, accusing Drake of breach of contract.

For Christiansen, his dome design was now precisely in the middle of a legal battle, with his integrity as an engineer put on trial. Spellman remained confident in the design team, stating, "We've got the world's best structural engineer [the Skilling firm]."[98] The *Seattle Times* reported: "John Christiansen, who designed the dome, and John Skilling . . . are 'very unhappy' with the hassle over the project and the roof. Christiansen said they are resolved to see it through. 'I have a great deal of pride in the design,' Christiansen said."[99]

Christiansen was "indignant" at the suggestion that his design was to blame.[100] With all the effort that had gone into the dome design, he found himself quibbling over inches with a builder that had no financial interest in continuing the project. All the while, the integrity of Christiansen's work was now viewed skeptically in the public eye.

King County Multipurpose Stadium, Seattle. Aerial view, with twelve segments cast, 1975.

Luckily for Christiansen, the county moved quickly. Once Drake stopped work on the job, and made it clear that the company had no intent to return, the county reached out to other contractors to finish building the stadium. On December 10, 1974, Spellman recommended the Peter Kiewit Sons Company to complete the construction. Beginning work on January 20, 1975, Kiewit was able to successfully separate the formwork from the concrete dome with little effort, and construction resumed. Kiewit added additional work shifts to make up for lost time, and by April, twelve roof segments were cast, and by June, over twenty-five of the forty segments were complete.

Just as the dome was nearing completion, the stadium was officially anointed with its name. On June 27, 1975, the *Seattle Times* announced: "Stadium has a name—the Kingdome."[101] The

King County Multipurpose Stadium, Seattle. (*above*) Aerial view, with twenty segments (half of the total) cast, 1975. (*left*) View from the top, nearing completion, 1975.

name effectively branded the Seattle stadium alongside other multipurpose arenas around the country, like the Houston Astrodome and the Louisiana Superdome. The name had been commonly used as a shortened version of the "King County Domed Stadium," and sportswriter Georg N. Meyers had called for an "official sanction" of the name as early as 1973.[102] Once selected, however, the name stuck and became synonymous with Seattle sports.

On July 18, 1975, the final segment was poured, and on July 22, the King County Stadium roof became completely self-supporting. Based on original estimates, the roof was completed almost two months ahead of schedule.[103] The King County Stadium had a grand opening on March 27, 1976, with over fifty-four thousand people. Spellman reinforced the public process that led to the Kingdome: "I believe . . . the stadium represents the very spirit of the people who voted for the original bond issue and then had the patience and determination to see it through. In an era when so many projects are either delayed to death or abandoned because there isn't a perfect consensus as to how to proceed, the stadium stands as a symbol that we can get something done."[104]

King County Multipurpose Stadium, Seattle. Complete exterior view, 1976.

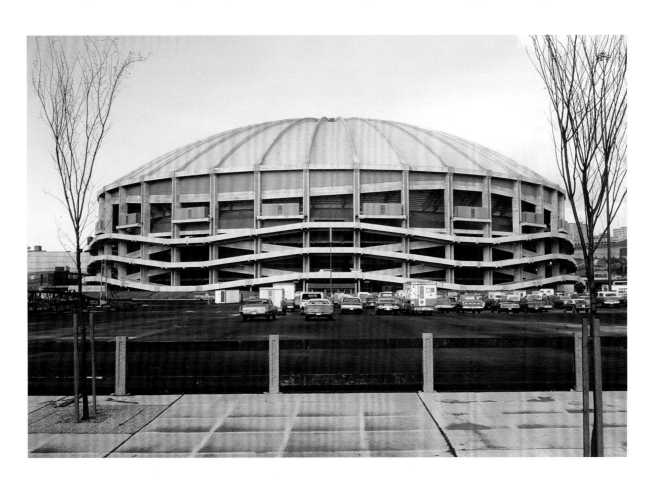

Architecturally, the Kingdome was celebrated by those who understood the intense budget scrutiny that had produced the building and who appreciated its austerity as a symbol of its economic design. Architectural critic Patrick Douglas wrote: "Functional, straightforward and honest, it is a practical what-you-see-is-what-you-get kind of edifice, a paean to the Scandinavian sensibility of a city where high praise is a shrug of the shoulders and a 'could be worse.'" He added: "It is a blood-and-guts sports arena pure and simple, a fitting addition to a city built by longshoremen and lumberjacks."[105]

Others in Seattle saw the Kingdome as the modern monument of the time. The sports journalist Royal Brougham celebrated the monumental yet public nature of the Kingdome: "There it stands, the most magnificent and imposing sports and convention center in the nation. The multi-purpose amphitheater dwarfs the Roman Coliseum and Babylon's Temple of Artemis. . . . It is a monument to man's daring imagination, ingenuity and intelligence. It is the largest indoor 'people's place' in the West, uniquely awesome in size and inspiring in its beauty. And it is owned by we the people!"[106]

Princeton professor David Billington (1927–2018), author of several publications and active with the IASS, had followed Christiansen's work for some time. In comparing the Kingdome to the work of Maillart and Nervi, Billington was critical of the exterior of the Kingdome, which "dominated its locale by contrast" and seemed "to fall short of its visual potential."[107] But Billington admired the "overwhelming" dramatic interior volume as "appropriate to the much larger space in the American work." For Billington, the Kingdome was most significant for its scale: "Thus it is a prototypical engineering design whose major significance for the future of thin shell concrete roofs is its demonstration that works of structural art can be built at the largest scale and with a

Kingdome interior, ca. 1977

relatively limited budget so long as engineering designers strive to achieve efficiency, utility and aesthetic quality in all their works."[108]

In 1978, Billington stated that the Kingdome was "one of the most successful thin shell concrete structures in the United States, if not the world."[109] Despite all the challenges the project faced, the economy of the design remained one of its most celebrated attributes. The county reported: "Inflation and unanticipated delays have escalated the cost of the stadium and its projected completion price is now estimated at $60 million—a remarkably low figure in view of the nation's spiraling economy."[110]

The county intended to recover some of these costs through its lawsuit with Drake, and the litigation surrounding the stadium would go on for many years. The county hired William Dwyer as its lawyer, while Drake hired Stuart Oles—two well-known Seattle attorneys who engaged in a heated trial. Christiansen spent two days on the witness stand, defending his design. After months of debate, on July 7, 1978, the judge ruled in favor of the county

Sports fans, Kingdome parking lot, ca. 1978

and ordered Drake to pay the county $12.2 million including interest. It was a victory for the county and a further vindication for Christiansen.

The Kingdome became a signature building on the Seattle skyline. Once completed, it immediately fulfilled its promise of hosting professional football, professional baseball, and professional soccer—franchises that Spellman had acquired in the meantime. It also served as a true multipurpose venue for holding concerts, performances, and exhibitions, providing an indoor space scaled to the sporting and cultural ambitions of Seattle.

Jack Christiansen's design talent definitively played a role in the execution of the Seattle Kingdome—rivaling those of any architect involved. In many ways, Christiansen's ability to adapt his structural design to the complex project demands throughout the process is what made the Kingdome a reality. Through his efficient, cost-effective design, Christiansen gave the city of Seattle's its first large-scale professional stadium.

The Kingdome also firmly connected Christiansen with the legacy of structural design throughout the world. He was asked to speak on the Kingdome in the decades that followed and took great pride in the entire design process. Christiansen later stated, "The Kingdome was my symphony . . . the chance to highlight the virtues of concrete in all its uses—the way that concrete should be used."[111] The Kingdome used a "combination of cast-in-place and precast concrete, thin shell concrete, and conventionally reinforced and prestressed concrete—all in different ways, appropriate to their nature."[112]

CHAPTER 7

RETIREMENT AND PERSEVERANCE

Christiansen's Late Career, 1976–2000

C OMPLETED IN 1976, the Kingdome was the apex of Christiansen's career in thin shell concrete, but the conditions in the surrounding industry were continuing to change. Thin shell concrete began decreasing in popularity across the world, replaced by other, more open-air structural systems. Responding to this trend, Christiansen branched out— designing with tension-cable roofs and steel space-frame structures in similarly creative ways. In 1983, interested in leaving the complexities of a large firm behind, Christiansen took an early retirement and became an independent structural engineer. Smaller-scale work from this point forward reveals Christiansen's continued devotion to thin shell concrete—a passion demonstrated through his adamant, but ultimately futile, opposition to the Kingdome demolition.

DECLINE OF CONCRETE SHELLS

Despite the completion of the Kingdome in 1976—globally recognized as the largest freestanding concrete dome in the world—the future of thin shell concrete was hotly debated. As early as 1969, Alfred L. Parme, the author of the original ASCE Manual 31 for design of cylindrical shells, claimed that the "mass market for shells is fading" and "the popularity of shells has not attained the promise it gave a few years ago."[1] Parme attributed this change to the tendency of owners to want to minimize labor costs and speed up construction, inherently favoring mass-produced building components that could be quickly assembled on-site.

In April 1970, the American Concrete Institute (ACI) hosted the Symposium on Concrete Thin Shells, organized by ACI Committee 334, with a "state-of-the-art survey of design, analysis and construction of thin shells" and a particular focus on the United States. It was the first major shells colloquium in the United States since the world conference in 1962. At an ACI conference, Anton Tedesko delivered the keynote lecture "Shells 1970—History and Outlook," in which he acknowledged a decline in the widespread use of shells.[2] He claimed that because engineers were forced to accept lower design fees, with no financial incentive to produce more cost-effective structures, they were selecting simpler, prefabricated structures for typical building projects. Still, Tedesko saw a bright future for shells:

> A new generation of shell designers has grown up in the United States. They add the gift of imaginative, artistic sensibility, and greater computer technology to the practical knowledge, understanding, and intuition of the older generation. These younger men are more structural sculptors than mass-producers of shells, more partners in the architectural teams than engineer-workers producing, as in former days, an inexpensive structural system to carry the architect's loads. Shell design has been advanced as a result of this development.
>
> Rationally-designed shell structures will have the greatest promise whenever there is cross fertilization between the minds of the designer and the constructor.[3]

Referencing computer analysis, architectural partnerships, and construction integration, Tedesko's statement encapsulates Christiansen's design approach. Although Tedesko does not mention Christiansen by name, he included an image of his International Exhibition Facility in New Orleans as his final graphic, with the caption "[The] first small translational barrel shells were built in Chicago (designed by the author), thirty years later Skilling-Helle-Christiansen-Robertson designed these eye-catching translational shells of vastly greater magnitude."[4] Christiansen also presented a paper on the International Exhibition Facility at this conference.[5]

Despite this optimistic outlook from Tedesko, the global engineering research community also began moving away from thin shell concrete structures. Other structural systems such as cable-grid structures, pneumatic (inflatable) structures, space frames, and prismatic structures promised new definitions of material efficiency, lightness, and transparency with much lower labor costs. Recognizing this trend, the International Association of Shell Structures, founded by thin shell concrete pioneer Eduardo Torroja, amended its own name to become the International Association for Shell and Spatial Structures (although still maintaining the initials IASS) in August 1971. The additional term *spatial* was defined to include a wide range of other structural systems.[6] The air-supported roofs of both the US Pavilion at Expo '70 in Osaka, Japan, and Frei Otto's 1972 Munich Olympic Stadium signaled exciting new directions in efficient, expressive structural designs.[7]

The 1970s also saw many changes in the architectural community. Throughout the 1950s and 1960s, thin shell designs had aligned with the prevailing architectural priorities of lightness and material efficiency. But the 1970s brought a significant shift in architectural ideas. Several factors, including the rise of environmental concerns and the OPEC embargo in 1973, changed the way architects thought about operational energy.[8] The rise of postmodern thinking changed the aesthetics of architectural representations and generally placed less emphasis on structural form.[9] As a unity of structural and architectural form, thin shell concrete was simply not as attractive as it once was.

DECLINING WORK WITH MAURY PROCTOR

Christiansen experienced this decline in "mass-market" shells through his work with Maury Proctor. Proctor was finding that his all-concrete building option was becoming less and less cost competitive, with warehouse buildings shifting toward the use of more pre-engineered, prefabricated components. Tilt-up concrete walls remained a popular option for warehouse enclosure, but light-framed steel or wood-frame roof structures were less expensive and significantly easier to construct on-site than the thin shell concrete umbrellas. Proctor and Christiansen maintained that their all-concrete building cost less over its life span than other roof systems, by having lower insurance costs, and attempted to highlight this fact in their advertisements.

Proctor had placed classified listings in the *Seattle Times* since 1967, under the title "Industrial Property," offering "ALL-CONCRETE BUILDINGS at low cost." In 1972, Proctor began adjusting the title of the listing to highlight the nature of his system, adding "Pre-engineered-Fireproof" before "All-Concrete Buildings."[10] In 1973, he paid for an advertisement to run twice (in January and March): "At last! A fireproof all-concrete building at low cost." Proctor highlighted "no interior firewalls or sprinklers to obtain the lowest fire insurance rates" and "spans to 60 feet."[11]

Emphasizing cost savings for industrial buildings, Proctor ran several ads in fall 1973.[12] In December, a large advertisement he ran included a mail-in coupon for more information and showed an image of the Tri-Cities Airport (completed in 1967). The advertisement attempted to take aim at the competing construction types: "This all-concrete building offers you low cost UMBRELLA COVERAGE for your total business investment, and it's the *only* one that does."[13] Proctor even tried developing a reusable form system for cylindrical shells—in a return to the form widely used fifteen years before. However, these promotional efforts did not return sufficient business. Proctor began changing the title of the classified listing, attempting to appeal to different markets, but in May 1974, the listing ran for the last time.[14]

Proctor had built over thirty projects between 1960 and 1972.[15] In total, Proctor and Christiansen had designed and constructed over 1.5 million square feet of inverted umbrella shell roofs.[16] However, he built just one warehouse in 1973, two in 1974, and one last Shell Forms Inc. building in 1977 (for Sundstrand Data Control in Redmond, Washington).

With dwindling demand, Proctor transitioned his business away from building construction into the manufacturing of high-end mechanical components for a variety of other industries. The machine shop that had been used to create the steel and fiberglass formwork for hyperbolic paraboloid shells was gradually retooled with more advanced equipment (drill presses, lathes, etc.) for higher-precision fabrication in aluminum or other metal alloys. In 1973, Proctor Products began advertising for welders and metal fabricators in the *Seattle Times*.[17] Proctor began servicing several of the aerospace companies in the Puget Sound region, in a return to his previous career in aviation. Christiansen

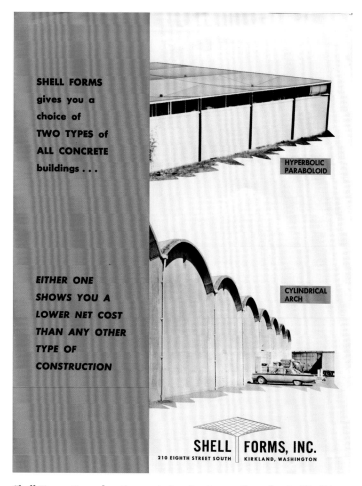

Shell Forms Inc. advertisement showing two options for shell buildings, hyperbolic paraboloid and cylindrical "arch," ca. 1974

stayed in touch with Proctor, but by 1978, the business arrangement with Shell Forms Inc. was over.

A NEW ERA OF STRUCTURAL DESIGNS

Christiansen's designs for long-span structures began to change as well, as he embraced the use of other structural systems to fit specific project needs. Matching architectural and economic demands, Christiansen branched out and began to use systems such as cable nets and space-frame roofs with the same creativity and attention to detail as before, but meeting different architectural intents.

In 1972, Christiansen designed Temple Beth El with Minoru Yamasaki in Bloomfield Hills, Michigan. In designing for one of the oldest Jewish congregations in metropolitan Detroit, Yamasaki moved away from his interest in precast concrete, adopting a different formal and structural strategy. Yamasaki designed a tent-like structure, recalling the earliest synagogue forms, and sought to achieve lightness through pure tension—not thin shell behavior.[18]

The temple consisted of a cast-in-place concrete base and towers that served as the frame for a cable-hung roof. Two sets of sloping end columns supported curved beams that ran along the top of the temple. Their curve created an opening between the two beams, covered with a skylight, creating an oculus-like quality. The cables sloped downward from these curved beams to a perimeter edge of concrete. A copper roof enclosed the space and covered the cables. The building marked a transition for Yamasaki—away from the intricate, delicate ribs of concrete to smoother, flowing forms.

EXPO '74

Christiansen's work with Yamasaki set the stage for another tension-cable structure, for the US Pavilion at the World Exposition—Expo '74—held in Spokane, Washington, in 1974. In contrast to the "World of Tomorrow" theme of the 1962 Seattle fair, Expo '74 was the first environmentally themed world's fair, with the slogan "Celebrating Tomorrow's Fresh New Environment."[19] Emblematic of the emerging design orientation within architecture, the fair had no interest in creating the utopian designs of the future, but rather was more sensitive to environmental concerns. "The Earth does not belong to man, man belongs to the Earth" became the fair's catchphrase.[20]

With this change in perspective came a change in approach to the structures built for the fair. The fairgrounds were located on Havermale Island in the middle of the Spokane River. The most prominent building of the fair was the US Pavilion, with its motto "Man and Nature: One and Indivisible." The form of the pavilion was specified to respond to the theme of the expo and provide a "temporary, economical cover of contemporary design for a large exhibit area."[21] After the close of the expo, the designers were asked to leave only an attractive landscaped area.

The temporary nature of the structure directed the designers away from a permanent, thin shell concrete option. The specified design objectives similarly reflected changing design priorities, including conservation not only of materials, cost, and time but also of energy. In December 1971, the Department of Commerce issued a request for proposals for the US Pavilion and selected a design team in June 1972. However, by December 1972, the General Services Administration determined that the design and fee proposals substantially exceeded the specified budget.[22] In January 1973, the design was awarded to the architectural firm of Naramore, Bain, Brady & Johanson (NBBJ)—only fifteen months before the fair was to open. With William Bain Jr. taking the design lead, the firm brought in Jack Christiansen as the lead structural engineer to address the specific requirements of the pavilion and do so on an extremely tight schedule.

US Pavilion, Expo '74, Spokane. Model view, 1973.

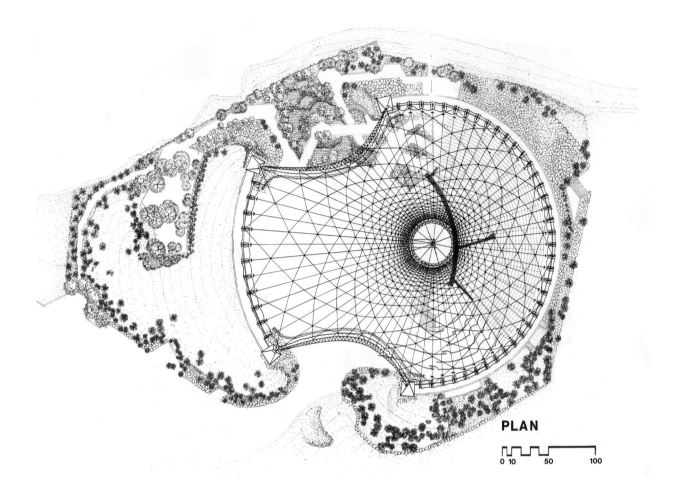

PLAN

0 10   50   100

US Pavilion, Expo '74, Spokane. Structural plan showing projection of overhead cable arrangement, 1973.

As with Temple Beth El, Christiansen developed a stressed cable structure that resembled a giant tent, with grass and trees underneath. The structure consisted of a large translucent roof of coated fabric covering a network of pretensioned bridge cables. The cables, suspended from a central 50-foot-diameter crown ring and a 150-foot-high trussed-steel mast, were anchored to perimeter concrete piers with steel-truss arches spanning 145 feet on each side for pedestrian circulation and views.

The large cable network defined the profile of the pavilion's roof. It consisted of one set of primary, radial cables draping down from the crown ring and two sets of secondary cables spiraling upward with an arching geometry.[23] With this arrangement, the initial tensioning of the primary cables created tension in the secondary system, allowing the different individual cables to define a single surface geometry. The jacking tension force varied from 30,000 to 50,000 pounds in the primary cables and was 25,000 pounds in the secondary cables, effectively giving each cable a compressive capacity.

Once the cables were connected at their intersection points, their opposing curvature created a rigid structural surface—a triangulated space frame with great strength and stiffness. In this arrangement, Christiansen used the same geometrical principle of double curvature in the cable network as he did with his concrete shells, and he referred to the pavilion as having a "soft shell roof." The cable network was covered with a translucent, polyvinyl chloride (PVC) fabric that performed no structural role.

For the structural analysis, Christiansen used a combination of computer and physical model tests. He and the architects at NBBJ built a scale model of the structure, hung weights from different locations, and measured the changes in form.[24] This model load test resulted in only slight changes to the final geometry. The computer analysis verified the geometrical stability of the form. With a double-curvature geometry, the structure was inherently stable, and with only lightweight covering, the cable network retained plenty of reserve capacity. Ground was broken for the project in May 1973, and it was completed by May 1974.[25]

The final report on the pavilion celebrated its "natural" appearance as "an expression of environmental concern. . . . The structure's smooth, graceful contour harmonized fully with the surrounding shoreline terrain. The trans-

US Pavilion, Expo '74, Spokane. Cable supported structure before covering, 1973.

lucent fabric roof allowed natural light to penetrate the Pavilion by day and interior lights to transmit soft glow from within at night."[26]

In 1974, the *Seattle Post-Intelligencer* published a profile of Christiansen. Celebrating the US Pavilion, the article described the structure as having the "lines of a trillium blossom," an example of Christiansen's "artistic design and engineering precision." Christiansen said of his guiding principles: "Know-how plus imagination are the necessary ingredients for the successful creativity that makes a building or a bridge stand out above all others in design and beauty."[27]

The entire structure was intended to be temporary, and Christiansen saw the life span of the PVC to be no more than seven years. This temporary nature contrasted with Christiansen's orientation toward the long-lasting, durable nature of thin shell concrete. He questioned the structure's susceptibility to rupture in adverse weather conditions or corrosion over time. He also questioned the high cost of the woven steel cables and synthetic fabric structures.[28] Christiansen enjoyed the design experience, and the opportunity to explore a different structural system, but remained unconvinced that tension-cable roofs should replace thin shell concrete.

US Pavilion, Spokane. Completed view, fair in foreground, 1974.

Christiansen designed with other types of structural systems during this time as well. In 1980, he designed the Museum of Flight in Seattle with the architect Ibsen Nelsen (1919–2001).[29] The museum was commissioned to detail the history of aviation and highlight Boeing's innovation in aircraft design. Having designed several long-span airplane hangars in thin shell concrete, Christiansen was used to the expansive needs of housing aircraft, but, adapting to the changing times, he would need to design a structural system that was drastically different.

The building, with over 136,000 square feet, had to fit on a 7-acre site adjacent to Boeing Field.[30] While there was plenty of room to expand on the site, the building had to keep a low roof profile to stay below the Being Field flight path requirements. Mimicking the trajectory of an airplane taking off, the building would have a gently sloping roof plane, requiring a planar structural system—effectively ruling out any thin shell concrete option.

Nelsen also wanted the building to be nearly transparent to its surroundings. Part of the exhibit display called for full-size aircraft to be hung from the interior of a long-span, 50,000-square-foot "grand gallery."[31] These aircraft were intended to be visible from the exterior—to give the impression of flight, as a flyby. As many as twenty full-size aircraft—including a DC-3 and a B-17—would be suspended from the roof in several different locations.

Museum of Flight, Seattle. Segment of modular space frame lifted for assembly, 1986.

Christiansen had to develop a roof system that could support such heavy loads, in a variety of different locations, while keeping the structure open.

Christiansen designed the structure as a space frame—a three-directional trussed framework of steel—enclosed by glass panels. The roof, trapezoidal in plan, was 380 feet long on the eastern side and 170 feet on the western side, with a 182-foot span in between. The roof would be supported by perimeter walls of trussed, tube steel columns. Several proprietary space-frame structures were available at the time, as patented systems.[32] The project team considered using one of these systems, but Christiansen thought he could design a more effective system for less cost.

Christiansen applied the same modular, segmental thinking he used to construct thin shell structures to the steel space frame. The entire roof was divided into identical, modular pyramids, triangular in plan. Roughly 19 feet on a side, each module consisted of a triangle of 10-inch-deep steel beams on top and three 5-inch-diameter steel pipes connecting downward from each corner to form an inverted pyramid. The resulting assembly was roughly 11 feet tall. Three of these pyramids were then bolted together at their corners, creating a planar roof surface. With each triangle defining a 60-degree angle, the three modules alternated with three triangular openings of the same size around the connection point. Smaller, 3-inch-diameter steel pipes connected the low point of each pyramid below.

Museum of Flight, Seattle. Completed interior view, 1987.

Each pyramidal module could be easily fabricated at ground level and combined in any multiple of three into larger panels. These panels were supported by the exterior wall, with only two temporary support towers needed to erect the entire roof. The complete roof became a two-way spanning system, meaning that the entire assembly could be supported only by the perimeter walls, with no interior columns. The roof also worked as a rigid roof diaphragm, able to resist the high wind forces collected by the 75-foot-high glass walls. The building was required to meet the strict performance standards of the National Energy Code. The large amounts of window area reduced the need for interior lighting, with triple-layered glazing to reduce solar gain.[33]

The space frame that Christiansen designed was bid competitively with other space-frame systems and selected on a low-cost basis. Christiansen claimed that his system was constructed at only half the cost of competing, proprietary space frames. Construction began in summer 1985, and the museum opened in July 1987.[34] The local media described the building as a "crystal lattice," "glowing with natural light," with "30 gleaming airplanes suspended from a glass ceiling in defiance of gravity, seeming to roar forward in rainbow-colored configuration."[35]

Christiansen appreciated these new directions in structural design and found he could design with them creatively. He understood the advantages that space-frame structures offered, but was discouraged by the high material cost of high-strength steel and expert labor needed for fabrication and erection. As with tension-cable roofs, Christiansen was still not convinced that they provided a clear benefit over concrete shells.

Christiansen remained a firm believer in thin shell concrete and was dedicated to continuing to prove its viability, in the face of critiques. In April 1979, David Billington invited Christiansen to Princeton University to present the past twenty-five years of his work in

Museum of Flight, Seattle. Completed exterior wall panel, 1987.

Museum of Flight, Seattle. Completed aerial view, 1987.

thin shell concrete.[36] Christiansen also delivered a second lecture at Princeton, in which he directly compared thin shell concrete to other long-span options.[37] While the profession was transitioning away from a widespread application of thin shell concrete, Christiansen continued to seek out individual projects well suited to thin shell concrete.

ROYAL SAUDI NAVAL STADIUM

Christiansen designed one, large-scale thin shell structure in the late 1970s—outside the United States—working with NBBJ on a grandstand structure for the Saudi Arabian Royal Navy. Begun in the late 1970s, but not completed until the mid-1980s, the stadium facility was designed to accommodate international-class soccer, Olympic track-and-field events, field hockey, equestrian events, and outdoor gymnastics. The Saudi Arabian government wanted a dramatic grandstand structure, and concrete was more readily available than steel in the Middle East at the time. In addition, the primary function of the grandstand was to provide shade for the viewing audience—with no need to enclose (or environmentally condition) large volumes of space. The stadium grandstand seated seven thousand spectators, with

Royal Saudi Naval Stadium, Jubail, Saudi Arabia. Side view of grandstand support and cantilevered overhead shell, ca. 1985.

three hundred VIP theater-type seats in a glass-walled, fully air-conditioned enclosure—suitable for the Saudi royal family.

As a result, the entire structure was concrete, with a dramatic cantilevered, thin shell canopy. Twelve large vertical piers rose from the ground, supporting a cantilevered middle and upper level of seating. The piers continued up, and each supported a thin shell roof slab that cantilevered out 75 feet from the face of the piers to provide sun, wind, and rain protection for the seating below. This repetitive system—twelve piers and twelve canopies—allowed for substantial reuse of forms in the casting of the 2.5-inch-thick shell.

To achieve the dramatic cantilever, Christiansen carefully incorporated post-tensioning cables within the thickness of the roof slab. This required precise alignment of the cables, with only slight thickening of the shell, to keep the induced force contained. The upper- and middle-tier support girders were also post-tensioned, and the tread and riser framing and concourse framing were another combination of precast, prestressed, and cast-in-place concrete.[38] Exterior finishes were concrete, with color derived from various aggregates and with texture in selected areas.

Despite the completion of this large-scale grandstand, the debate surrounding thin shell concrete continued. In 1982, David Billington published a second edition of his *Thin Shell Concrete Structures*. In the preface, Billington acknowledged that "some have assumed during the past few years that thin shells are going out of fashion," attributing this position to three

Royal Saudi Naval Stadium, Jubail, Saudi Arabia. Completed stadium from a distance, ca. 1985.

reasons: analysis and design of concrete shells were too complex and expensive, their forms were too expensive to build in high-labor economies, and they "visually [had] lost their general appeal to the general public and to the architect."[39] Billington directly disputed this position: "While we could easily argue against each of these reasons, the most stunning refutation of them all stands in the works of a few contemporary designers like Heinz Isler and John Christiansen, whose recent designs clearly overcome any general objections to thin shells." Isler had successfully completed over 250 shell structures in Switzerland, a country with a cost of building similar to the United States', by developing his own design and construction processes. For Billington, and others in the engineering community, Isler and Christiansen still indicated a promising future for concrete shells.

MORE CLIMBING AND TRAVEL

Within a shifting design landscape, Christiansen continued to center himself through climbing excursions in the outdoors. In May 1980, Christiansen and his wife climbed Mount Adams in southern Washington. After weeks of seismic activity surrounding nearby Mount St. Helens, Christiansen knew that an eruption was likely, but did not anticipate an event. However, once he was atop Mount Adams, with a clear view of Mount St. Helens, the eruption began. A large plume of hot ash blew out of the volcano, giving Christiansen a front-row seat. He quickly took several photographs of the expanding

(*above, left, opposite*) Eruption of Mount St. Helens as seen from Mount Adams, May 18, 1980. Suzanne and Jack Christiansen in foreground.

cloud. The images were later published in *National Geographic*, with the caption "An eyewitness to the unimaginable, Suzanne Christiansen drops awestruck on Mount Adams as the blast, 35 miles away, quickly spreads to a 20-mile halo of death."[40] A climbing partner took another photograph of Christiansen and his wife.

At the same time, Christiansen began to travel internationally more and more. The Christiansen family traveled to Spain and Portugal in 1973, East Africa in 1974, Egypt in 1975, Great Britain in 1977, and Southeast Asia in 1978. These trips were again oriented around architectural monuments and climbing excursions. Christiansen visited Stonehenge, the Pyramids of Giza, Durham Cathedral, and castles throughout England. In 1980, he took another significant trip to western Europe, visiting Brunelleschi's dome once again in Florence and traveling to Rome for the first time. He visited the Pantheon, photographing the large interior volume and massive dome. He noted the details of the rib structure, with a similar system of radial ribs as he had used on the Kingdome. But Christiansen was most fascinated by the Roman Coliseum, taking many different photos of its structural forms and interior space. He also photographed exhibit diagrams of its possible construction, and use, applying his engineering thinking to this historical monument.

DEPARTURE FROM THE FIRM

For Christiansen, in the early 1980s, the demands at the firm were becoming more and more complex. Skilling, Helle, Christiansen & Robertson had expanded into a nationally significant structural engineering firm, with both

Seattle and New York offices flush with large-scale projects and new opportunities. John Skilling maintained strong relationships with architectural clients in Seattle—including NBBJ, Callison Architecture, and the Richardson Associates (TRA). In 1981, Skilling had aligned with the developer Martin Selig and architect Chester Lindsey to design the seventy-six-story Columbia Center, the tallest building on the West Coast upon completion. Leslie E. Robertson remained in charge of the New York office, with a growing list of architectural clients of his own, including Philip Johnson and Gunnar Birkerts, and increasing independence from the Seattle office. Robertson was in the process of designing the seventy-two-story Bank of China Tower in Hong Kong, with architect I. M. Pei. As the firm grew, Christiansen was pulled between the two offices and found himself doing more managing and less designing.

During a partner meeting in summer 1982, Skilling and Robertson both remarked how energized they felt by the expanding scope of work. Despite being in their fifties, they agreed that they could never see themselves retiring, and, indeed, both men would continue to work for many more years.

Christiansen had a different perspective. While he enjoyed the design work immensely, he did not always enjoy the logistics of a large-scale firm. Christiansen commented that being a partner in a firm was a lot like being in a marriage, requiring a close, personal relationship with the other partners. These relationships required constant communication and frequent compromises that took his energy away from his structural designs. He envisioned a new mode of practice on his own, with smaller, more intimate designs and increased independence and flexibility. In summer 1982, Christiansen decided to exercise an option for early retirement and, by January 1983, left the firm at the age of fifty-five.

Christiansen's departure spurred a major reorganization of the engineering firm. With Christiansen leaving, Skilling and Robertson decided to formally separate the Seattle and New York City offices. Robertson founded Leslie E. Robertson Associates (LERA), and Skilling reorganized and renamed the Seattle office, recognizing the next generation of leaders, to become Skilling, Ward, Rogers & Barkshire. Both firms continued to design landmark structures around the world.

CHRISTIANSEN'S INDIVIDUAL PRACTICE

Upon leaving the Skilling firm, Christiansen agreed to a five-year separation deal with Skilling and Robertson, with a non-compete clause and assurances that he would help finish projects that were already under way. This allowed

Christiansen to complete the Museum of Flight project and a Baltimore Convention Center. In addition, he took a teaching job as an affiliate faculty member at the University of Washington, jointly in the Department of Architecture and Urban Planning and the Department of Civil Engineering.[41] He oversaw a senior design project for a high-rise building for engineering students. He was also invited to speak on the Kingdome at Purdue University and the University of Wisconsin–Madison.[42] While Christiansen enjoyed this academic role for several years, he was not passionate about becoming an educator full-time and, in 1986, started his own practice as an independent structural design engineer.

Working as an individual, Christiansen saw his career reduced in scale (in terms of the size and number of projects he designed), but he was able to be more selective in the types of projects he pursued. Also, with fewer projects and less administrative responsibility, Christiansen could be more involved in all parts of a project—from design through construction.

He was also free to pursue thin shell concrete structures on his own. Despite a diminishing interest from the architectural community, Christiansen strongly believed in the advantages that shell designs offered, particularly in long-span applications, and he continued to lobby for their use. In 1986, he presented his past work at the "Space Structures for Sports Buildings" conference held in Beijing, where he used his 1962 hangar at Boeing Field and the Kingdome to show that hyperbolic paraboloid, thin shell concrete was "most suitable" for the long-span roofs required for sports buildings.[43]

He became chairman of the joint ACI-ASCE Committee 334 on concrete shell design and construction. In 1987, he oversaw a session at the ACI fall convention in Seattle, dedicated to presenting the state of the art of hyperbolic paraboloid shells. Presenters included William C. Schnobrich, Alex C. Scordelis, Mark Ketchum, and Milo Ketchum, along with Christiansen. Their presentations covered analysis techniques, specific shell configurations, and details of construction and were published as a collection in 1988.[44] In the introduction to the volume, Christiansen wrote: "It is hoped the information presented in this publication will encourage the design and application of this efficient and economical type of construction."[45]

He also presented a paper titled "Hyperbolic Paraboloid Performance and Cost," in which he provided simplified calculations for approximating the design of hyperbolic paraboloid shells. He offered simple ratios, combining quantities of height, length, depth, and thickness as generalized geometrical criteria. Christiansen hoped these rules of thumb would help other engineers use hyperbolic paraboloid shells, in a similar way that ASCE Manual 31 had helped him design cylindrical shells at the beginning of his career.

As validation of his efforts, Christiansen continued to find applications for his thin shell concrete designs. In 1989, he was contacted about designing another multipurpose stadium facility, to be located on the Central Washington State Fairgrounds in Yakima. The operational success of the Seattle Kingdome, which was continually hosting professional sports and exhibitions, had inspired the construction of another (smaller) facility that could serve eastern Washington in a similar manner.

Christiansen began working with the Seattle firm LMN—founded by architects George Loschky, Jud Marquardt, and John Nesholm, who departed NBBJ in 1979—and the local architects at Loofburrow. With Christiansen as the structural designer, the 7,000-seat SunDome became a miniature version of the Seattle Kingdome. The primary open space was covered with a thin shell concrete dome, spanning 270 feet—less than half the span of the Kingdome—and had a rise of 40 feet.[46] The dome was composed of twenty-four triangular, saddle-shaped segments, 3 inches thick, with a similar pattern of stiffening ribs.[47] As in Seattle, the perimeter tension ring of post-tensioned concrete contained the thrust generated from the dome.

SunDome, Yakima, WA. Shell segment geometry, 1989.

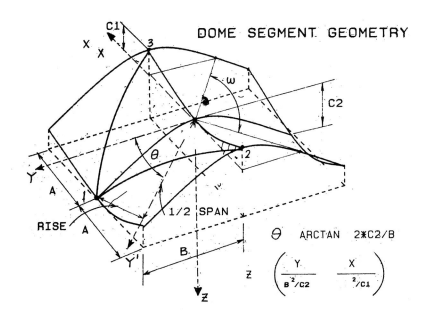

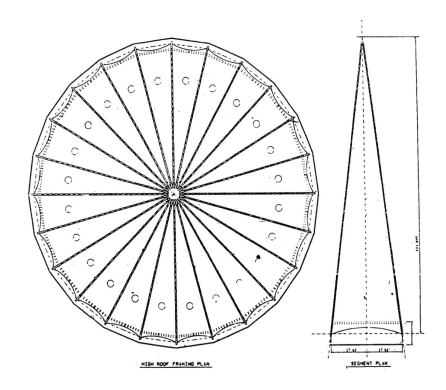

HIGH ROOF FRAMING PLAN   SEGMENT PLAN

SunDome, Yakima, WA.
(*above*) Wedge segment detail
for shell roof, 1989.
(*right*) Under construction,
1989.

This design facilitated a similar rotating, reuse of forms for construction, as used on the Kingdome, except on a smaller scale. Christiansen designed a falsework system that required only two wedge-shaped forms, supported by a central tower and a perimeter track—just as for the Kingdome. The contractor, however, decided to use a more conventional system of aluminum

SunDome, Yakima, WA. Complete, 1989. (*above*) Exterior. (*left*) Interior.

scaffolding supporting six, wedge-shaped forms. Still able to take advantage of the repetitive use of forms, the contractor poured concrete on each one four times to complete the twenty-four-segment dome.

Unlike the Kingdome, the SunDome was surrounded by auxiliary spaces enclosed in rectangular, concrete block walls, for a total of 90,000 square feet. These spaces contained ticket booths, restroom facilities, and meeting rooms—giving the SunDome variously scaled spaces for different activities. The SunDome was profitable in its first year of operation and remains a successful event space. For Christiansen, the SunDome was evidence that thin shell concrete still had much to offer.

An article in *Concrete Construction* profiled the economics of the Sun-Dome's construction costs and benefits.[48] Christiansen contributed a sidebar titled "Concrete vs. Fabric Domes," in which he argued for the continued use of thin shell concrete. He stated the advantages of concrete in economy, fire resistance, and durability and as a locally sourced material. Based on his own research, he compared the cost per square foot of each, claiming that the SunDome roof cost $14 per square foot, while the average cost of a fabric roof (either air-supported or cable-truss) was more than $30 per square foot.[49]

BAINBRIDGE ISLAND HIGH SCHOOL GRANDSTAND

In 1991, Christiansen proposed the design of a 1,500-seat, covered grandstand for a high school athletic field in his home community of Bainbridge Island. For this design, Christiansen had to directly compete on a cost basis with standardized wood and steel grandstands that the school district was also considering. Christiansen took this small-scale project as a personal challenge to prove that thin shell concrete not only was the low-cost solution but could also create an interesting structure.

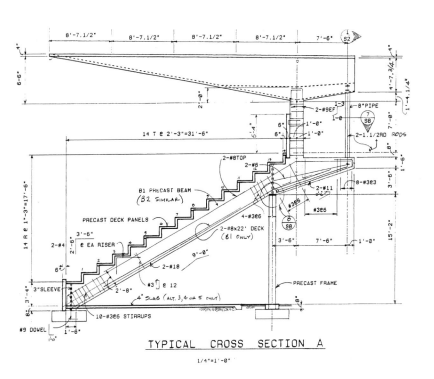

Bainbridge Island High School Grandstand, Bainbridge Island, WA. Typical cross-section of grandstand and overhead canopy, 1991.

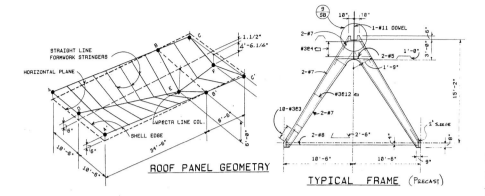

ROOF PANEL GEOMETRY

TYPICAL FRAME (Precast)

Bainbridge Island High School Grandstand, Bainbridge Island, WA. Overhead canopy and support detail, 1991.

To meet this challenge, Christiansen used nearly all the construction techniques he had learned in his career. He would make the grandstand a modular, site-cast, precast concrete structure that required a minimal amount of easily constructed formwork and a minimal amount of material. He used overhead hyperbolic paraboloid shells (21 feet wide) to cantilever 34.5 feet over fourteen rows of grandstand seating. Each shell was only 2 inches thick. These shells also extended 8.5 feet behind the support and were fastened down with tension ties to a precast concrete support structure. This same structure, a single precast concrete element, was also shaped to the supported overhead canopy and the angled grandstand seating. This support structure rested on inverted V frames along the backside of the grandstand, which stabilized the framework and provided lateral strength and stiffness.

This module of overhead shell, support structure, and inverted V frame was repeated nine times to create the overall grandstand structure. Precast tread and riser beams with aluminum seats spanned in between. With such simple compositional elements and repetition that allowed for the reuse of formwork, the grandstand became relatively simple to build.

All the concrete elements were cast at the construction site. The contractor could build wood forms at the ground level, and with the hyperbolic paraboloid shells, support structure, and inverted V frames easily defined by straight lumber, the forms were inexpensive to create. Concrete could be poured in a conventional manner, without expensive material handling equipment (e.g., a concrete pump). Once cast, these simple elements could be erected with a single mobile crane in just a few days.[50]

Christiansen saw the Bainbridge Island Grandstand as emblematic of his "economy-minded" approach to structural design, with "nothing willy-nilly."[51]

Bainbridge Island High School Grandstand, Bainbridge Island, WA. (*top*) Casting of canopy shells on-site, 1991. (*bottom*) Erection of precast pieces, 1991.

Bainbridge Island High School Grandstand, Bainbridge Island, WA. Complete, 1991.

The economy of concrete shells was the driver of his architectural aesthetic. He continued to see thin shell concrete as the most efficient structural form, and the Bainbridge Island Grandstand remained one of Christiansen's favorite works.

Christiansen cited the Yakima SunDome and the Bainbridge Island Grandstand as examples of the continued viability of thin shell concrete. In a 1990 article, Christiansen mentioned these structures to show "how hyperbolic paraboloid concrete shells may be used to construct a variety of efficient, economical and attractive roof structures."[52] Perhaps responding to the broader decline of shell construction, Christiansen used the article to argue for the continued advantages of thin shell concrete.

Christiansen reiterated the benefits of shell construction and detailed the total cost savings that his thin shell concrete structures provided to building owners over the life of the structures. With changing labor rates, Christiansen acknowledged that thin shell concrete buildings did not always have a lower construction cost than other wood or metal buildings (though that was the case with the grandstand). Still, other factors made shell structures a viable option. He included insurance rate quotes that showed overall costs were still lower for a fireproof, all-concrete building. He argued that the durability of concrete to resist high humidity and extreme temperature variations was unmatched by other structural materials.

Directing his comments toward clients, he argued that architectural, mechanical, or electrical equipment could be easily attached to the shell at any location and that the interiors were "free of dust-collecting surfaces or ledges." He also added: "Many shell configurations of concrete shells are visually attractive."

In this article, Christiansen was attempting to convince other designers to take up thin shell construction. While before, he was in a position to build a large number of shells himself—either through projects at the Skilling firm or through Shell Forms Inc.—now he was more interested in sharing and disseminating his knowledge to the larger building community. The article contained detailed drawings of shell geometry and specific reinforcing information, providing the basis for others to continue this work. Christiansen concisely summarized the most important aspects of constructing shells, as a guide for others to follow: "The key to their economy is their high level of structural efficiency, their use of small quantities of low-cost materials, and the reduction of otherwise high formwork costs by using straight-line-defined and moveable forms." Despite this article and Christiansen's involvement with the ACI-ASCE Concrete Shell Design and Construction Committee, very few concrete shell structures were being built in the manner Christiansen described.

THE DEMISE OF THE KINGDOME

In Christiansen's late career, he continued to climb significant peaks nearly every weekend. In 1989, the *Seattle Post-Intelligencer* profiled Christiansen, not as an engineer, but as a mountain climber. Highlighting his prolific climbing record, the article noted that Christiansen was approaching one hundred peaks climbed in the Olympic Mountains. When asked why he embarked on such a "sylvan" mission, Christiansen replied: "The Olympics are my backyard and I like a challenge."[53] A similar article appeared in the *Bainbridge Island Review*.[54]

But this time in Christiansen's life is also marked by the demise of the Seattle Kingdome. The Kingdome had served Seattle as a multipurpose venue for many years, but in the end, the same economy of construction and material efficiency that made the Kingdome possible would ultimately

Downtown Seattle and waterfront, 1977

become the justification for its downfall. In the 1980s and 1990s, Christiansen became a kind of spokesperson for the Kingdome, taking a more vocal role than any other architect, contractor, or public official involved, speaking not only to the structure but to the building as a whole.

The controversy had started early. Less than five years after its completion, the Kingdome began facing questions of aesthetics, rather than function, and it was Christiansen who provided the most passionate response. In 1981, Christiansen delivered a presentation to the ASCE, titled "Aesthetic Choices Leading to the Kingdome."[55] Christiansen said that while the Kingdome was the only multipurpose stadium that made a profit and that nearly "all opposition to the Kingdome has vanished," the "only criticism in recent years has been from visiting aesthetic critics from the east who spend less than 24 hours in Seattle and usually don't even enter the building."[56]

Christiansen defended the appearance of the Kingdome, not as an intentionally composed work of art, but as simply a product of the "functional and technical requirements of the building," where "severe funding restraints did not allow a single penny to be spent for 'aesthetic design elements.'" He stated: "Although it can be said there were no purely aesthetic choices, nevertheless, at every step of the way, there were decisions and choices made that were both functional and economical, *and* aesthetic."

Christiansen proceeded to analyze the Kingdome's appearance according to a list of design elements and principles, taken from the 1969 book *Art Today*.[57] He discussed the circular shape of the Kingdome, which accommodated a variety of seating arrangements; the roof dome, which provided clearance for sports below; the tension ring (a horizontal transition between the base and roof); the columns, ramp system, and mechanical units; and the texture and color of materials—all as essential structural elements that became part of the building's aesthetic. Christiansen concluded: "Given this rather concise statement of art, I believe the Kingdome can be treated as a work of art, even though no purely aesthetic choices were made and no theories or philosophies of aesthetics were applied. The Kingdome is imperfect in many respects, and its site and surroundings are less than ideal. However, taken as an organized, coordinated whole, with a potential durability rivaling that of the pyramids and the Colliseo of Roman times, the Kingdome may prove to be a significant structural art form."[58]

In April 1982, the *Seattle Times* featured the Kingdome as "the Northwest's most spectacular example of long-span design."[59] Christiansen was quoted throughout the article, highlighting the building's durability: "There would be no way to build a more durable building. The roof structure is a very efficient shape and appropriate structural form. Not all long-span roofs have been such. . . . There is too much building cheap, tearing down and building again. It gives me a good feeling to think about people coming to see the Kingdome in a few thousand years, like the Coliseum in Rome." Christiansen's reference to the Roman Coliseum recalls the earlier comparison by Royal Brougham and Christiansen's own 1980 trip. The statement also reinforces how Christiansen had begun to see the Kingdome through a historical lens, in context with other large-scale works.

But as the 1990s approached, the life span of the Kingdome came directly into question. While functional, the Kingdome design did not capture the affections of the entire city. With little design time (or budget) devoted to integrating the building into the surrounding neighborhood, the Kingdome was an eyesore to many. With its gray concrete exterior, others commented that the Kingdome looked unfinished even when construction was complete. While the dome provided shelter from the rain, the humidity and air quality of the interior environment was at times uncomfortable. Operationally, the building left much to be desired as well. Spectators complained about a lack of bathrooms and how much time it took to reach their seats via the long-winding access ramps.

In addition, a series of faulty waterproofing membranes on the dome's exterior had led to water intrusion and staining of the dome's interior surface. This cosmetic issue began to create a dingy interior aesthetic, but soon became even more serious. This leakage caused some of the large acoustical ceiling tiles to become saturated with water (to a weight of over 20 pounds each), and on July 19, 1994, just before a Seattle Mariners baseball game, several tiles fell more than 200 feet to the arena floor. Out of concern for public safety, this event prompted outcry from the community and closure of the Kingdome for repairs.

The tiles had been fastened to the concrete with sheet-metal ties, cast into the shell from underneath when the dome was poured. When Christiansen went to assess the fallen panels, he saw that the sheet-metal ties were all flattened down and had not been embedded in the concrete at all. Christiansen speculated that the workers had stepped on the ties during construction and that only the adhesion between the surface of the panels and the concrete was holding them in place. With the security of the remaining panels uncertain, the county decided that all forty thousand panels, covering the entire interior volume, were to be removed. With difficult access to the dome surface, the expensive replacement of the interior panels and installation of a new waterproofing membrane placed a $70 million financial burden on the King County taxpayers. Frustrations with the Kingdome were growing.[60]

Much to Christiansen's disbelief, some in the local media claimed the dome itself was falling in—while there was never any structural concern with the roof. In August 1994, Timothy Egan, writing in the *New York Times*, claimed that the Kingdome had "become a grunge incarnation of every homeowner's worst nightmare. The roof leaks. The ceiling is falling. And what was once merely ugly is now unsightly to the point of revulsion."[61] Egan fundamentally questioned the domed enclosure as a sports facility, and whether baseball should be played "indoors under cement skies," criticizing the fake grass and humidity of the interior environment.

These operational troubles with the Kingdome coincided with changes in professional sports. In 1992, the Baltimore Orioles opened Camden Yards—an intimate, open-air, baseball-only stadium that was designed to match the character of the surrounding neighborhood. A new generation of baseball facilities was soon to follow, mimicking the heritage of baseball stadiums of the 1920s. Steel-truss structures provided more openness to the outside conditions, and exposed brick walls referred back to earlier stadiums such as Fenway Park in Boston and Wrigley Field in Chicago. More situated within urban contexts, new baseball stadiums in Denver (Coors Field, 1994) and Arlington, Texas (Globe Life Park, 1995), were part of a new breed of intimate, baseball-specific stadiums.

These new stadiums were built to match the changing financial model of professional sports rooted in public-private corporate partnerships. The growing influence of sports on American culture resulted in larger television contracts and escalating player salaries. Team owners began demanding more control over their facilities as revenue-generating entities. The multipurpose arenas built and controlled by municipalities were becoming a hindrance to the new model of professional sports. The era of the domed, multipurpose arena was ending.

In the 1990s, new structural systems continued to displace thin shell concrete in these new stadium proposals. In 1989, the steel-framed Toronto SkyDome opened, showing off the world's first, large-scale, rigid, retractable roof.[62] In 1992, the Georgia Dome showcased a long-span "tensegric" structure—a lightweight, steel cable, and fabric supported dome. Throughout the 1990s, the IASS meetings focused on lightweight structures, with topics such as tensegrity, lattice structures, and deployable structures. Concrete shells retained a presence at the meetings, but discussions were relegated to more industrial applications (e.g., cooling towers), rather than architectural structures.[63]

In this changing landscape of professional sports and engineering communities, the Kingdome, less than twenty years old, no longer fit the image of a modern arena. With mounting public unhappiness, a movement for a new outdoor baseball stadium began in Seattle. The new stadium would emphasize baseball on natural grass, under open skies, as a return to the qualities of the past. An open facility also had no need for the expensive and difficult-to-maintain heating and cooling systems of the Kingdome. It also promised both views of downtown Seattle and the waterfront and a more intimate fan experience—a sharp contrast to the contained Kingdome experience. To accommodate these two goals, in Seattle's wet environment, the baseball team proposed a retractable roof—an "umbrella" for rainy days.

But public opinion was not unanimous on the particulars of a new baseball stadium. On September 9, 1995, a King County proposition to raise the sales tax by 0.01 percent to fund the projected $240 million cost of a retractable roof stadium and repairs for the Kingdome failed by a small margin. Voters were skeptical of new taxes, especially considering the maintenance costs already invested in the Kingdome.

With this rejection, discussions of moving the Mariners to the Washington, DC, area immediately began, adding familiar urgency to the stadium dispute. Following a spirited playoff run by the Mariners, the state legislature intervened to keep the team in Seattle. The legislature authorized a different funding package, one that did not require a public vote, where new construction would be supported by a new food and beverage tax, a rental car surcharge, and other revenue streams. In October 1995, the state legislature sent the issue back to the King County Council, and after a short deliberation, the council voted in favor of allocating over $389 million in public funds. Combined with $126 million provided by the Mariners organization, the total cost of the new stadium, later named Safeco Field, would be $517 million.[64] For comparison, the roughly $67 million spent to build the Kingdome in 1976 (a figure slightly higher than the $60 million quoted by Spellman) equated to only $162 million in 1997 dollars—over three times less.[65] The baseball-only Safeco Field was to be a high-tech spectacle—with a retractable steel-truss roof, surrounded by brick facades. Built directly south of the Kingdome, it opened in 1999. Symptomatic of a larger trend, the Kingdome was one of five stadiums closing to Major League Baseball in 1999, including Tiger Stadium in Detroit, the Milwaukee County Stadium, Candlestick Park in San Francisco, and the Houston Astrodome.

Even with the construction of Safeco Field, the Kingdome was intended to remain the primary facility for NFL football. But larger changes were tak-

The Seattle Kingdome and recently completed Safeco Field, 1999

ing place throughout the league as well. Team owners were beginning to realize the value of high-end stadium facilities. Allegiances between cities and teams were tested on financial terms, and team owners began direct negotiations with different municipalities across the United States, resulting in the relocation of several franchises. In November 1995, the Cleveland Browns signed a deal to relocate to Baltimore. After the 1996 season, the Oilers were scheduled to leave Houston (and the Astrodome) for Nashville, Tennessee, after promises of a state-of-the-art stadium. Both the Raiders and the Rams had recently left Los Angeles, for Oakland and St. Louis, respectively, with stadium concerns at the center of the disputes.

In this financial landscape, Seahawks owner Ken Behring became more vocal in his displeasure with the Kingdome. Sensing an opening in the California market, Behring and his lawyer met with King County executive Gary Locke, to argue that the twenty-year-old Kingdome was already outdated and no longer suitable for professional football. With none of the revenue-generating amenities of the new stadiums being built, he stated: "There's no income, no advertising, no club seating. We can't compete with other teams. We can't compete."[66]

Behring tried to amplify his complaints by claiming that the Kingdome had severe structural deficiencies, making the building unsafe in the event of earthquakes. In 1995, the county had commissioned a survey of the structural safety of the Kingdome, to assess its ability to resist newly understood seismic concerns.[67] Christiansen, long retired from practice, was not involved in the assessment.

The engineers at RSP/EQE, led by seismic expert John Hooper, had performed new computer analyses on the Kingdome.[68] The results showed the robust capacity of Christiansen's design and the overall safety of the building. The report identified only minimal deficiencies in interior structural columns (supporting only grandstand seating) and recommended a retrofit through a fiberglass wrapping of columns. The report stated that the improvements were "minor to moderate" in nature.[69]

Despite this assessment, Behring claimed that the report indicated there were larger structural issues with the Kingdome. Both parties agreed that the Kingdome needed renovation, but disagreed on the scale of the work needed. The county maintained that minor seismic deficiencies could be corrected with roughly $15 million of work, with Behring claiming that the number was more like $150 million. Without a new facility, or significant county-funded renovation, Behring would relocate the team.

Forcing the issue, on February 2, 1996, Behring announced that he was loading up the team's equipment vans for a move to Anaheim, California.

With his questions surrounding the Kingdome, Behring was banking on league approval of his relocation despite no formal agreement of a stadium deal in southern California. To resolve the issue, the NFL owners called a collection of Seattle's seismic experts to a meeting, to hear about the structural safety of the Kingdome firsthand. On March 13, 1996, this group, including Hooper and Jon Magnusson, CEO of Christiansen's former firm, now named Skilling, Ward, Magnusson & Barkshire, successfully convinced the NFL owners that the Kingdome remained a safe venue for professional football.

With the safety of the building confirmed, Behring no longer had sufficient cause to press for an immediate move to California. Instead, the NFL opposed Behring's hubris and threatened a fine of $500,000 plus another $50,000 for each week the team operations remained outside of the Northwest. By late March, Behring had brought the team back to Seattle, but his long-term intentions with the team were now clear. With declining influence in the Northwest and the Seahawks committed to the Kingdome through 2005, Behring decided to sell the team.

Anxious to find a buyer, King County located technology billionaire Paul Allen as a potential buyer. Allen was eager to keep the Seahawks in Seattle; however, he understood the changing model of professional football. Allen would agree to buy the team only if a new stadium, partially funded with public funds, was going to be built. Structural safety aside, Allen echoed Behring's claims that the Kingdome did not generate enough stadium revenue for him to have a financial interest in owning the Seahawks. Pro football teams share the ample revenues generated from television contracts and the sale of licensed merchandise with the professional league (the NFL), but the rest of a team's revenue comes from its stadium-generated sources, including large luxury boxes, retail spaces, and concession infrastructure. Built in a previous era, the Kingdome simply didn't have enough amenities.

Allen claimed that the Seahawks were at the very bottom of the league in stadium revenue, and a new facility was needed to generate the revenue stream for decades to come. Renovation was rejected as too costly, and suburban sites for a new stadium were turned down because they lacked the civic infrastructure already in place near the Kingdome. Allen's preference was to raze the Kingdome and build a twenty-first-century version of Husky Stadium—the open-air, grandstand stadium at the University of Washington.[70]

A large public debate ensued about whether the Kingdome could operate without a professional football team, in the event that the team built another stadium somewhere else. The *Seattle Times* reported: "Politicians, pundits and people talking football at the neighborhood tavern are grumbling about plans to destroy the building. . . . It's a bass note in the debate over whether to destroy the dome to make way for the $402 million, open-air football stadium that billionaire Paul Allen wants before he'll buy the Seahawks."[71]

Allen gave lawmakers an ultimatum: ask voters to fund a new stadium or stand by the Kingdome and hope for the best with Ken Behring. Allen developed statewide television ads, hired topflight lobbyists, and set up an internet page complete with a calculator to add up the Kingdome debt, all claiming that the Kingdome was not economically feasible for football.

Led by Washington State governor Gary Locke's stadium task force, the state legislature passed House Bill 2192 with legislation enabling a new public funding package. Referendum 48 would raise $325 million of taxpayer money to finance the demolition of the Kingdome and the construction of a new football-specific stadium, exhibit center, and parking garage. With the passage of this referendum, Allen would agree to purchase the Seahawks and contribute $100 million to the new stadium construction. The referendum would be placed on the June 17, 1997, ballot for voter approval.

Others in Seattle voiced their support for the Kingdome. Bill Sears, retired Kingdome sales and promotion manager, said the Kingdome was always considered one of the best-operated stadiums in the country, pulling in $1 million in profit its first year. "The building is without a question the most functional building in the country," he said. "That, I guess, is part of the tragedy. If it were a building that were so-so and it didn't have that type of reputation, then you'd say, well maybe it has outlived its use. But it's hard for me to reconcile that."[72]

Observing the ongoing debate in the local media, Jack Christiansen inserted himself into the Kingdome debate. On June 3, 1997, Christiansen wrote an opinion piece to the *Columbian* (Vancouver) newspaper, addressing several issues that Allen's referendum proposed. Christiansen also focused on the continued utility of the Kingdome:

> The Kingdome can be remodeled to satisfy the reasonable demands of pro sports teams. More luxury suites and club seats plus luxury lounges, more restrooms and concession stands and attractive glass-enclosed entrances and exterior screen walls can be added for a cost in the $100 million range.
>
> The all-concrete Kingdome structure with its record-size concrete dome could have a 1,000-year life. It is ironic that, were the Kingdome to avoid the threatened demolition and reach an age of 50 years, it would almost certainly attain landmark status and be preserved for its significant historical value. However, the power and money of and pressure from big-time professional sports, with their

huge salaries and escalating franchise values, are such that citizens and their representatives are coerced into building new facilities with tax dollars.[73]

Christiansen's claims about the Kingdome's historic status were not unfounded. On June 15, 1997, the *Seattle Times* published a letter to the editor from Princeton professor David Billington:

> As a structural engineer and historian of technology, I am writing to support as strongly as possible the popular movement to save the Kingdome from demolition. This great structure continues to be of great benefit to Seattle economically and it is the most impressive concrete-roof structure in the world. There is no doubt that it will be a national historical landmark comparable to the Brooklyn Bridge in New York.
>
> The Kingdome is by far the most economical large-scale fixed covered stadium ever built. It operates in the black and cost about $\frac{1}{20}$th the amount paid for the Montreal Olympic Stadium, the only other concrete stadium of comparable size.
>
> The Kingdome is also the most visually dramatic roof structure of any large covered stadium anywhere, both from above and from the inside. It is an extraordinarily striking thin-shell design featured in textbooks that proves how the very best large-scale works of modern American civilization can (and indeed must) combine efficiency, economy and imagination.
>
> The Kingdome will certainly be a national historical landmark when it reaches the necessary age for qualification (50 years). I have worked with the National Park Service to develop criteria for designating large structures as national historical landmarks and the Kingdome will clearly fit such criteria.
>
> The criticism of the Kingdome recalls the criticism of the Eiffel Tower just over a century ago when it was also threatened with removal. Fortunately, popular support helped save the Tower for Paris, where it has since become its primary symbol. The Kingdome embodies the ideals of efficient, economical and imaginative design that our civilization needs to carry into the 21st century. Great works of modern engineering help define our society. Seattle has such a work and I hope that the people of the city will vote no in order to save this great work.

Despite these impassioned letters, voters narrowly approved Referendum 48 on June 17, 1997. Paul Allen agreed to purchase the Seahawks and keep the team in Seattle. Officially, the Public Stadium Authority was then only conducting a site-selection process for the new football/soccer stadium. But barring some unforeseen challenge, the Kingdome site was nearly certain to be chosen at the end of the process.[74]

With this slim window of opportunity, Christiansen sprang out of retirement to try to rescue the Kingdome. Recognizing the changing model of professional football, Christiansen suggested alternative sites for a new stadium and created several designs to renovate the Kingdome as a complimentary parking garage and convention center—adding new floors at the 100 level for concerts and sports; inserting new offices, shops, and entrances; and creating large windows for natural light. He "inundated the [Public Stadium Authority] with hand-sketched designs and calculations to illustrate his ideas and bolster his case."[75]

Through these designs, Christiansen provided a reuse of the massive interior space: "The Kingdome could then be used for sports events, conventions and parking. The dome could be divided horizontally, with the ground floor used for parking and a new upper floor or floors for sports or conventions. He estimated the remodeling would cost roughly the same as building a new convention facility and tearing down the dome."[76]

Christiansen was not alone in these efforts. Construction consultant Ray Biggs and Ford Kiene, chairman of the Kingdome's advisory board, were among a dozen professionals who donated time and skills to brainstorm a new version of the Kingdome. Their solutions, they claimed, would not only be useful for the region but also be affordable for taxpayers and enable the Kingdome to retire its own debt in a timely fashion.[77]

Failing to find traction with the public, Christiansen eventually hired a lawyer, Vince Larson. Larson argued that there was a case to be made that the new stadium's environmental impact statement did not adequately address the issues Christiansen raised. Christiansen remarked: "It's not my thing to raise a fuss and deal with public issues, but I feel so strongly a terrible mistake is being made."[78] Christiansen could not make sense of the economics of the new stadium proposal and why it demanded the demolition of the Kingdome:

> Having reviewed all the data, I can't understand how a conclusion
> was reached to demolish the building. Somebody seemed to think
> that the Kingdome is a very ordinary thing you throw away if the

numbers don't come out a certain way. I take issue with the numbers and the process.

Studies have also reported the Kingdome needs some seismic upgrades and other capital improvements, which have been estimated at $34 million. But even if that figure is accepted at face value, the total of about $110 million is much less than half the $325 million public share of Allen's new stadium.[79]

Christiansen's understanding of history and world travels reinforced his sense of how historic the Kingdome was: "Huge things like this have lots of potential uses. It never occurred to me it wouldn't be here for a very long time. A lot of other structures were. What about the Los Angeles Coliseum or the Rose Bowl and those facilities? They've been there for 70 or 80 years. There is no reason why they shouldn't stay."[80]

Despite all the appeals and outrage, in March 1998, the Public Stadium Authority decided to demolish the Kingdome and build a new, open-air football stadium on the same King Street site. The decision was not universally shared. Former King County executive John Spellman also questioned the demolition of the stadium. When asked about the inevitable loss of the building, Spellman stated: "Will I shed a tear? Probably. We live in a very disposable society. We throw things away awfully fast. It was built with all the best

The implosion of the Seattle Kingdome, March 26, 2000

advice available. I don't know if that sense has gone away, but times have changed. The facility is still what it was in the beginning. It has not deteriorated; it's not dilapidated. Nothing's wrong with the Kingdome."[81]

Spellman also reminded people that there wouldn't be a need for a new basketball arena, a new baseball field, and a new football stadium if there hadn't been a Kingdome to bring teams and fans and fame to Seattle. "They wouldn't have anything to see if it weren't for the Kingdome," he said. "Let that be its epitaph."[82]

The demolition of the Kingdome had as much to do with the changing culture of Seattle and the American sports landscape as it did with the individual building. The era of the expansive, multipurpose stadium had passed. The *New York Times* reported: "America's domed stadiums, which were once called technological marvels, but after barely a generation, are increasingly considered obsolete and antiseptic by fans and owners."[83]

DEMOLITION DAY

With plans for a new football-only stadium on the Kingdome site, the Kingdome was dramatically imploded on March 26, 2000. The demolition was broadcast on five local and one national television stations. After the evacuation of several city blocks and meters were set up to measure seismic resonance, the demolition company drilled six thousand holes in the concrete structure, filled the holes with explosives, and detonated them. In a few seconds, the Kingdome was reduced to a remarkably minimal pile of rubble— resembling the shell of a cracked egg.

Jack Christiansen took the demolition personally. The demolition company called Christiansen for advice on how to best demolish the concrete structure, but he had no interest in assisting. Christiansen was offered front-row seats to the demolition show, but he forcefully declined. In discussing the demolition, Christiansen stumbled for words: "It's maddening to think they could spend millions of dollars to tear it down. They're not going to get one bit of help from me. I feel terrible. Sick. Mad as hell, actually."[84]

Christiansen had envisioned the Kingdome lasting for a thousand years, yet it lasted less than twenty-five. And with no structural problems, the building could have continued to stand in service to the community for many years to come. Christiansen remarked: "This is the dumbest thing I have ever seen done. A perfectly good, usable building, a structure that could have lasted 1,000 years, an engineering marvel, and they blew it apart. For what? It happened because big-time professional sports is out of

The crushed concrete shell of the Seattle Kingdome, March 26, 2000

control, that's why. They get to do anything they want. This facility stood in the way."[85]

Christiansen's wife, Sue, lamenting the lack of maintenance, said: "They never painted it like they were supposed to. They took from it but never gave back."[86] Christiansen's opposition to the Kingdome demolition reflected his design values as much as any of his constructive work. He believed in the long-term durability of his built work and the conservation of materials, and he saw the demolition as a shortsighted accommodation of changing demands and a waste of a structure. He appreciated the stripped-down aesthetic of the Kingdome, as a display of modesty, restraint, frugality, and responsibility.

The day after the implosion, the *Seattle Times* reported on the conflicting sentiments within the city. While some claimed that the Kingdome was a cold, unhospitable venue with poor acoustics, problematic sight lines, and not enough bathrooms, others celebrated its "symmetry and strength" and its true multipurpose nature. While some said that they would not miss the gray, imposing concrete form, others felt it was a building that "reflected the times" and an artifact of simpler times in professional sports.[87]

The demise of the Kingdome was also emblematic of the changing architectural trends. Shell structures became popular because of their perceived ability to merge structural and architectural form, best suited to simple enclosures of space. As the Kingdome demonstrated, buildings in the late 1990s were facing far more complicated issues. The enclosure of a volume was becoming secondary to issues of energy, environmental impacts, and finance. An enclosed, weatherproof dome had been an essential part of the Kingdome's original design, thought to be a vital amenity for year-round sports in the Pacific Northwest. Public interest had also shifted back toward open-air, or retractably roofed, stadiums, even in the challenging climate of the Pacific Northwest.

To the end, Christiansen continued to believe that shell structures could offer the most beautiful structural solution. And for him, the Kingdome was the ultimate execution of this idea, as a creation of "sculpture on a grand scale": "I'll defy anybody to tell me it's not a beautiful roof. Anywhere you view it from, around the city, from flying into the city, it's a beautiful roof. You go look at these other roofs and they don't stack up."[88]

CONCLUSION

The Legacy and Future of Concrete Shells

DESPITE THE DEMOLITION of the Kingdome in 2000, Jack Christiansen maintained his faith in thin shell concrete through the end of his life. Following the broader decline of thin shell concrete in the 1990s and 2000s, more recent publications and academic activity have suggested a resurgence of interest in shell construction. Christiansen's receipt of the Eduardo Torroja Medal from the IASS in 2016 solidified his place within the global network of shell designers. Local preservation activity also shows a renewed appreciation for thin shell concrete and its role as a modern material in the Pacific Northwest. With this new enthusiasm, Christiansen's passing in 2017 evokes both the past triumphs of thin shell concrete and a new era of thin shells yet to come.

LATE BRIDGE DESIGNS

Christiansen continued to design in thin shell concrete into the 2000s, in long-span applications. In 2003, at the age of seventy-six, he completed one of the longest-spanning bridges of his career—the Centennial Bridge in Fairbanks, Alaska. The State of Alaska Department of Transportation and Public Facilities commissioned Christiansen to design a pedestrian-bicycle bridge to cross the Chena River, which runs through downtown Fairbanks. Made to commemorate the centennial of the city's founding, the bridge was required to span 180 feet across the river and connect to roadways on either side, giving the bridge a total length of 265 feet.

Christiansen considered many structural options for the bridge—including cable-stayed and suspension designs—but as before, he decided to use thin, post-tensioned concrete. The final design used an approach similar to Christiansen's earlier bridges in Seattle—a 13-foot-wide post-tensioned con-

Chena River Bridge, Fairbanks. Complete, view from the water, 2003.

crete bridge, with a 6-foot-deep triangular deck that reduced to roughly 3 feet in depth at the center. The bridge widened to 25 feet at the center, as a viewing platform. For the construction of the bridge, Christiansen designed a movable formwork system that allowed the bridge to be built in 15-foot sections, cantilevering from each side of the river and meeting in the middle—with no need for scaffolding or support in the river below.[1] Many local events, including the Yukon Quest dog sled race, start or finish at this bridge. Named the William Ransom Wood Centennial Bridge, the bridge was dedicated on September 5, 2003. Christiansen had used thin concrete once again for a cost-effective, materially efficient, long-span use.

THE FUTURE OF CONCRETE SHELLS

With Christiansen winding down his career, some reflected that Christiansen's work symbolized the end of an era and the conclusion of the global thin shell concrete movement. As the Kingdome's demolition approached, Princeton professor David Billington already viewed the Seattle stadium in historic terms, describing it as "the most dramatic roof structure of any large covered stadium anywhere" and the "culmination of a historic, world-wide movement."[2] Penn State architectural engineering professor Thomas Boothby echoed those remarks: "It was the last major thin-shell built in this country, and it's the biggest."[3] As a distinctly American designer in the late twentieth century, with the longest-spanning structures, Christiansen had secured a definitive role in the global history of shell structures in the modern period, but the future of thin shell concrete was less certain.

Christiansen certainly never wavered in his support of the medium, remaining steadfast in his dedication and envisioning a continued use into the future. The design values he held—of efficiency, economy, and the interrelationship between architecture and structures—were unswayed by public discussions in the press, political transitions, or the changing demands of professional sports. It was Christiansen's opinion that the near-universal qualities of thin shell concrete should continue to be valuable to architects, engineers, and builders.

Another major shell designer still practicing in the 2000s, Swiss engineer Heinz Isler, agreed with Christiansen's assessment. In 1999, Isler claimed that there were still opportunities for shells in modern architecture. At the time, Isler's office was producing between two and twelve shells per year in Switzerland and other countries, with labor conditions similar to those of the United States. He reiterated the importance of shells for their minimal use of material, maximum-bearing capacity in compression, and great simplicity as both structures and space enclosures—essential qualities. As improvements, Isler described how earth-covered roofs could make great use of thin shell concrete and insulation could be used as the permanent formwork for the shell. Isler's success was called the "Isler exception" within the IASS and seemed to prove that, in spite of global trends, concrete shells could still have a future as a viable construction option.

Other designers, elsewhere around the world, continued to have success with concrete shells. R. Sundaram (1935–), practicing in Bangalore, India, has continually designed with concrete shells since 1963, including a large number of long-span structures.[4] In 2010, Sundaram presented a paper on concrete shell roofs in Asia to an IASS symposium in Shanghai, concluding with the claim that concrete shells continue to have a bright future.[5]

As the 2000s progressed, the legacy of thin shell concrete in modern architecture—especially the work of Félix Candela, Eduardo Torroja, and Pier Luigi Nervi—continued to inspire speculation about whether concrete shells could once again emerge as an exciting construction type. In 2005, Columbia University professor Christian Meyer and practicing engineer Michael Sheer asked in *Concrete International*, "Do concrete shells deserve another look?"[6] Meyer and Sheer interviewed several engineers, architects, and academics—including Christiansen—on the reasons for concrete shells' decline and the factors hindering their resurgence, but the opinion they found was far from unanimous. While the majority agreed that fewer shells were being built, others contended that new innovative, yet proprietary, forming methods, like pneumatically inflated forms, were bringing forth a new era of concrete shells.

The authors found that most of those surveyed cited the high cost of construction due to the requirement of highly skilled labor—difficult to find in many construction markets—as a primary deterrent, precisely the issue Christiansen had dedicated his design practice to resolving. The authors also found that many believed that shells had simply lost favor with architects. Some stated that architects were less interested in structural economy as a driver of form and more interested in the extreme geometrical freedom offered by new design software, citing the work of Frank Gehry. As for the decline of long-span concrete shells for stadiums, some responded that the focus of the modern arena was no longer its structure and form but its ability to accommodate the economics of sports entertainment—an adequate summary of the Kingdome's demise.

The authors concluded that there was little consensus about the practical limitation of concrete shells. Concrete shells still offered the expressive, signature form and with proper management of the construction process could remain cost competitive with other systems. By embracing new concrete technologies and inspiring use by architectural designers, the authors optimistically speculated that "maybe great designers will revive a tradition that produced some of the most magnificent architectural landmarks of the 20th century."

The reason for the disappearance of shell structures continued to be a matter of debate. In 2015, UK architect Gabriel Tang echoed many of the earlier findings, arguing that through education, by understanding the construction methods used for shells, one may "bring about a renaissance of such structures" and enjoy the benefits of their "structural performance," cost efficiency, and "high aesthetic value."[7]

In recent years, shells of a variety of materials have shown signs of a resurgence. In 2006, the architect Toyo Ito designed the Meiso no Mori Municipal Funeral Hall in Kakamigahara, Japan, with a flowing, white, concrete shell roof—met with praise for its integration with the surrounding landscape.[8] Ito and structural engineer Mutsuro Sasaki used the undulations in the roof to house certain program elements, while allowing the roof to drain through the supporting columns—both techniques that Christiansen employed, through his hyperbolic paraboloid geometry. In the academic context, recently published textbooks have promoted shell design in a variety of materials, including timber, masonry, and high-strength polymers. In 2010, Edward Allen and a collection of engineers and educators published the textbook *Form and Forces: Designing Efficient, Expressive Structures.*[9] It promotes a revival of an earlier structural analysis method called graphic statics, which is particularly useful in the design of shell structures. In 2014, Sigrid Adriaenssens, Philippe Block, Diederik Veenendaal, and Chris Williams edited another educational guide, *Shell Structures for Architecture: Form Finding and Optimization.*[10] Both books provide significant assistance to architects and engineers in the conceptual and computational design of shell structures. The authors recognize the appealing mix of material efficiency and architectural expression that shells offer. Such utility with shell design, as presented through these texts, may indeed usher in the next generation of shell structures.

With this resurgence of interest, an understanding of Jack Christiansen's career offers valuable lessons and, possibly, practical tips on designing contemporary shells. As late as 2016, Christiansen continued to design with concrete shells in a speculative way. He produced drawings for the "Oval Dome," which would be better suited to the dimensions of a football stadium. With a span in the long dimension of 748 feet, the Oval Dome exceeded the Kingdome in interior volume. To cast the oval shape, Christiansen used two different forming geometries—wedge and trapezoid—unlike the single wedge for the circular Kingdome. Well versed in cost estimation, Christiansen also said that the design was significantly less costly than the structural systems used for other contemporary stadiums. His innovation in formwork was continuing to drive new structural forms.

In confirmation of his significant work and, perhaps, reflecting this renewed enthusiasm for thin shell concrete, Christiansen was awarded the Eduardo Torroja Medal in 2016 by the IASS. This award formally placed Christiansen within a global pantheon of creative engineers that included, among others, the award's namesake (Torroja), Heinz Isler, and Frei Otto. The award acknowledged Christiansen's role within a larger global movement of thin shell concrete, while celebrating the regional and temporal particularities of his work in the United States. While other Americans had been given the award, Christiansen was the first recognized for his built designs in thin shell concrete.[11] Presented at the 2016 conference in Tokyo, Christiansen's work was celebrated as a distinctly American contribution to the transnational conversation in thin shell concrete.

PRESERVATION

The preservation of many of Christiansen's buildings has mirrored the decline and then resurgence of interest in thin shell concrete structures. Several of Christiansen's signature works, including the Kingdome, were demolished. The Rivergate exhibition hall was demolished in 1995, before the Kingdome, to make way for a larger casino facility. The King County International Airport hangar was demolished in 2011, to accommodate a different runway orientation. In 2013, the elevated roadway of the Nalley Valley Viaduct was removed to expand the capacity of the freeway. In 2013, Mercercrest Junior High School was demolished to make way for a larger school building. In 2015, the Seattle School District Warehouse, Christiansen's first application of reusable formwork, was denied Seattle City Landmark status and demolished.

However, recent preservation advocacy through Docomomo WEWA (a local chapter of the international organization) has led to starkly different results. In 2017, three of Christiansen's projects—the Ingraham High School Gymnasium and Auditorium (NBBJ, 1959), a single-story office building for the Shannon & Wilson company (NBBJ, 1959), and a three-story office

Shannon & Wilson Office Building, Seattle. Completed 1959, designated Seattle City Landmark 2017.

Pacific Architect and Builder Inc., Seattle. Office Building. Completed 1959, designated Seattle City Landmark 2017.

building for Pacific Architect and Builder Inc. (A. O. Bumgardner, 1959)—achieved landmark status. The Landmarks Commission recognized Christiansen's role as a significant designer in the three buildings and his use of a distinct and innovative construction method. These buildings were also celebrated as part of the postwar expansion of the Northwest as a whole and Christiansen's role in shaping its distinctly modern character. Thin shell concrete had perhaps transitioned once again in the public eye.

OBITUARIES

When Jack Christiansen passed away, on August 16, 2017, his death was followed by an outpouring of admiration. Obituaries appeared in the local *Seattle Times* and *Bainbridge Island Review*, the regional *Daily Journal of Commerce*, as well as the national *Engineering News-Record* and *ASCE News*.[12] In each publication, colleagues and collaborators celebrated Christiansen's creativity in structural design, his role as a collaborator, and his significant impact on the Northwest.

His former partner Leslie E. Robertson recalled Christiansen's creativity with structural form: "Jack was able to take his ideas for structure and make an architectural form that jolted your eyes." Jud Marquardt, a former collaborator and founding partner of Seattle-based LMN Architects, remembered Christiansen as exploratory and always looking to enhance the form, constructability, and economy of every project. Marquardt welcomed Christiansen's creative contributions as an integral part of his designs: "If you knew Jack Christiansen was going to be your structural engineer, you would engage him in design strategy meetings at the outset."[13]

Christiansen's legacy of creativity in structural design lives on through the successor firm of Skilling, Helle, Christiansen & Robertson, now named Magnusson Klemencic Associates. Christiansen had mentored a young Jon Magnusson in the early 1980s, imparting the importance of collaboration and imagination when working as a consulting structural engineer. Magnusson recalled: "Jack was certainly one of the most creative engineers that I have ever met. When he worked with trace paper and a soft pencil on structural concepts, he actually created art."[14]

LEGACY

Jack Christiansen's work is a case study in the power of creative structural design. Throughout his career, he displayed a remarkable talent for developing innovative building forms—ones that were expressive and spatial, but also economical and structurally efficient. He embraced an approach to structural design that celebrated the economy of material, the permanence of his structures, and the total integration of engineering, architecture, and construction in the definition of building form.

With such an extensive body of work, spanning forty years, Christiansen fundamentally changed the built environment of the Puget Sound region, and thus his designs provide a valuable lens to understand the changing cultural conditions of Seattle and the Pacific Northwest. The rise and fall of Christiansen's work in thin shell concrete directly mirrors the transitions that the region experienced over this time. In the early 1950s, an extensive period of growth, the barrel-vaulted thin shell concrete structures were an exciting but cost-effective means of producing postwar buildings, offering new architectural possibilities. Later, working with the hyperbolic paraboloid, Christiansen helped expand the geometrical range of possible forms in modern architecture, reflecting the interest in exploring the structural aesthetics and spatial qualities of buildings. The 1962 World's Fair placed thin shell concrete at the center of Seattle's expanding national image, as a technical feat symbolic of the Space Age. As the 1960s continued, thin shell concrete became more of a structural form, praised for its cost efficiency and increasingly long-span capability. Repetitive warehouse structures and infrastructure projects continued to be built, as was the long-spanning Rivergate exhibition hall. During the course of the Kingdome project, taking place over many years, thin shell concrete construction transitioned from an exciting, expressive system to one of pure material and spatial efficiency. Thin shell concrete became a no-

Jack Christiansen in front of Bainbridge Island High School Grandstand, 2010

nonsense construction method, required by the times of financial austerity and used to simply enclose the most interior space with the least amount of material. The destruction of the Kingdome indicates how thin shell concrete had become a dated structural type in the public eye. Replaced by open-air, high-cost facilities, the enclosed Kingdome no longer matched the expanding characteristics and ambitions of the region in the twenty-first century. The resurgence of interest in concrete shells toward the end of Christiansen's life suggests, once again, the shifting of values and a potential future yet to come. Through the uniqueness of his design approach and the large number of structures he designed, Christiansen's work is an important part of modern architecture in the Pacific Northwest.

Though he worked collaboratively, with both architects and builders, the projects he worked on clearly reveal his own signature style as a structural engineer. Christiansen's work highlights the role that structural engineers play in the production of modern buildings. Christiansen demonstrated how, through creative structural contributions not only to safety and budget but also to aesthetics and construction processes, structural engineers can have an influential role. He was not a specialist in a specific project type, but displayed structural ingenuity in a variety of applications. He designed schools, churches, pavilions, airplane hangars, roadways, bridges, and even his own home, showing incredible versatility in his design range.

But Christiansen was most successful because of the personal connection he felt with his structures. Thin shell concrete was more than a structural type to Christiansen; it was the manifestation of his core design values—values he felt passionately about and maintained throughout his life. His dedication to the medium weathered significant changes in public opinion, only to find a potential revalidation in his final years.

As Christiansen's most significant design, the Kingdome remains a vital part of Seattle's urban history. Its likeness continues to be printed on T-shirts, hats, and other sporting paraphernalia. Relics of the Kingdome—including chunks of concrete collected after its implosion—are regularly found in online auctions. In 2015, fifteen years after the implosion, King County paid off the last of the Kingdome maintenance debt, sparking a wave of reflections and remembrances. In 2017, the Seattle Mariners celebrated their stadium history through a game-day giveaway promotion: a pair of snow globes, one featuring Safeco Field and the other the Kingdome. The gray mass of the Kingdome—articulated by the ribs of the roof and the surrounding ramps— is still instantly recognizable to many urbanites and sports fans in the Seattle area. The Kingdome lives on as a symbol of the beginning of professional sports in Seattle and a part of Seattle's physical transition into a major American city. The building served an essential purpose for King County and was instrumental in shaping the culture of Seattle. The same can be said of the work of Jack Christiansen.

APPENDIX

Jack Christiansen's Thin Shell Concrete Structures

The following list of thin shell concrete projects is drawn from Christiansen's own professional records. Architect Susan Boyle further expanded this list following her interview with Christiansen in 2009, and it has expanded significantly during the research undertaken for this book. This list is limited to Christiansen's work in thin shell concrete. As a partner in Skilling, Helle, Christiansen & Robertson for over thirty years, he contributed to many more projects, of different material and structural systems, which are not included here. This list provides the completed date of the project, the project name, best-known address, the associated architect (if any, if known), a brief description of the structural form (cylindrical shell, hyperbolic paraboloid, etc.), and accompanying notes. Notes indicate the projects where multiple shells were designed (i.e., high school gymnasiums and classroom buildings) and the current state of the structure (e.g., demolished, altered). Some entries remain incomplete but represent the best, currently available information on individual projects.

1954   **Green Lake Pool**
7201 East Green Lake Drive N
Seattle, WA
Daniel E. Lamont
Cylindrical (intermediate)

1955   **Ellensburg Senior High School Gymnasium, Workshop, and Music and Arts Building**
1508 E Third Avenue
Ellensburg, WA
John W. Maloney
Cylindrical (short and intermediate), multiple barrels with arched ribs
Demolished

1956   **Boeing Flight Test Hangar**
8987 Turner Street
Moses Lake, WA
NBBJ
Cylindrical, intermediate, multiple barrels with prestressed girders

1956 **Forest Ridge School Gymnasium**
1617 Interlaken Drive E
Seattle, WA
John W. Maloney
Cylindrical (long), multiple barrels with
edge beams

1956 **Nile Temple**
305 Harrison Street
Seattle, WA
Samuel G. Morrison
Cylindrical, multiple barrels

1956 **Seattle School District Warehouse**
1255 Harrison Street
Seattle, WA
John W. Maloney
Cylindrical (long), multiple barrels
Demolished

1956 **West Valley High School Gymnasium
and Classroom**
9800 Zier Road
Yakima, WA
John W. Maloney
Cylindrical (intermediate), multiple barrels
Demolished

1956 **Wilson Junior High School Gymnasium
and Cafeteria**
702 S Fortieth Avenue
Yakima, WA
John W. Maloney
Cylindrical (long), multiple barrels with pre-
stressed edge beams
First prestressed (post-tensioned) edge beams
in United States
Demolished

1957 **Asa Mercer Middle School**
1600 S Columbian Way
Seattle, WA
John W. Maloney
Cylindrical (long), multiple barrels

1957 **Bellevue High School Shop Building**
Bellevue, WA
Cylindrical (long), multiple barrels
Demolished

1957 **Highland Junior High School Gymnasium
and Classrooms**
15027 NE Bel-Red Road
Bellevue, WA
NBBJ
Cylindrical (short, long), multiple barrels with
arched, rigid frame stiffeners and edge
beams

1957 **Industrial Building**
Grand Mound, WA
Cylindrical (long), multiple barrels

1957 **Quincy High School Gymnasium**
16 Sixth Avenue SE
Quincy, WA
John W. Maloney
Cylindrical (long), multiple barrels

1957 **Wapato High School Gymnasium**
1001 S Wasco Avenue
Wapato, WA
John W. Maloney
Cylindrical (long), multiple barrels

1957 **Washington State Penitentiary Shop**
1313 N Thirteenth Avenue
Walla Walla, WA
John W. Maloney
Tilted cylindrical (north light), multiple shells

1957 **Wenatchee Junior High School
Classrooms, Gymnasium, and
Covered Walkways**
1620 Russell Street
Wenatchee, WA
NBBJ
Cylindrical, intermediate, multiple barrels,
hyperbolic paraboloid (inverted umbrella),
used as full shells and half shells

1957 **Yakima Valley Junior College Gymnasium**
S Sixteenth Avenue and Nob Hill Boulevard
Yakima, WA
John W. Maloney
Cylindrical (long), multiple barrels with
edge beams

1958   **Chief Sealth High School Theater,**
Classrooms, and Gymnasium
2600 SW Thistle Street
Seattle, WA
NBBJ
Cylindrical (short, intermediate, long), multiple
    barrels with arch stiffeners and edge beams

1958   **Hospital Canopy**
California
John W. Maloney
Conoid, double cantilever over arch stiffener

1958   **Hyster Warehouse**
5930 First Avenue S
Seattle, WA
Hyperbolic paraboloid (four-gabled), shells
    precast one on top of the other

1958   **Mercer Island High School**
Multipurpose Room
9100 SE Forty-Second Street
Mercer Island, WA
Bassetti & Morse
Hyperbolic paraboloid (saddle shapes), seven
    identical shapes joined to produce a dome
Demolished

1958   **Service Station**
Olympia, WA
Hyperbolic paraboloid (inverted umbrella),
    unsymmetrical double cantilever

1958   **St. Edward Church**
4212 S Mead Street
Seattle, WA
John W. Maloney
Cylindrical (long), single long barrel shell,
    Vierendeel girder edge beams

1958   **University of Washington Pedestrian**
Bridges
4124-4186 Montlake Boulevard NE
Seattle, WA

1959   **Ingraham High School Gymnasium and**
Auditorium
819 N 135th Street
Seattle, WA
NBBJ
Cylindrical (long), multiple barrels with pre-
    stressed shell, edge beam, and tied stiffener
    (the first known use in the United States),
    hyperbolic paraboloid (saddle shapes), three
    identical shapes joined

1959   **Pacific Architect and Builder Inc.**
1945 Yale Place E
Seattle, WA
A. O. Bumgardner
Hyperbolic paraboloid, gabled
First project with Maury Proctor

1959   **Shannon & Wilson Office Building**
1105 N Thirty-Eighth Street
Seattle, WA
NBBJ
Hyperbolic paraboloid, gabled

1959   **Western Washington University**
Gymnasium
189 Twenty-First Street
Bellingham, WA
Bassetti & Morse
Hyperbolic paraboloid, four shells

1960   **Bellevue High School Grandstand**
601-749 107th Avenue SE
Bellevue, WA
NBBJ
Hyperbolic paraboloid, cantilever

1960   **Mercercrest Junior High School**
8805 SE Fortieth Street
Mercer Island, WA
Bassetti & Morse
Hyperbolic, multiple/tied multipurpose room
    and umbrella
Demolished

1960   United Control Corporation Building
15001 NE Thirty-Sixth Street
Redmond, WA
Kirk, Wallace, McKinley & Associates
Hyperbolic paraboloid, multiple umbrellas
First use of Shell Forms Inc. system

1960   Wood Products Research House
220 130th Avenue SE
Bellevue, WA
Kirk, Wallace, McKinley & Associates
Hyperbolic paraboloid, wood shell

1961   Burlington Elementary
Burlington, WA
Hyperbolic paraboloid (inverted), multiple
    umbrellas

1961   Fine Arts Pavilion (World's Fair building)
301 Mercer Street
Seattle, WA
Kirk, Wallace, McKinley & Associates
Folded plate, multiple plates post-tensioned

1961   Washington Corrections Center
2321 W Dayton Airport Road
Shelton, WA
Bassetti & Morse (Walker & McGough, Curtis
    & Davis)
Hyperbolic paraboloid (four-gabled), large
    shells with prestressed ties

1962   Cedar Valley Elementary School Classroom
20469-20525 Fifty-Second Avenue W
Lynnwood, WA
Dan F. Miller
Combination of hyperbolic paraboloid shapes
Demolished

1962   Century 21 Entry Gates (World's Fair
building)
Bassetti & Morse
Hyperbolic paraboloid, freestanding wood shell
Demolished

1962   Church of the Good Shepherd
2310 128th Avenue SE
Bellevue, WA
Kirk, Wallace, McKinley & Associates
Diamond-shaped floor and wood roof shell

1962   International Pavilion (World's Fair
building)
155 Mercer Street
Seattle, WA
Walker & McGough
Hexagonal, modified hyperbolic paraboloids
    with single supporting column
Demolished

1962   King County International Airport Hangar
7277 Perimeter Road S
Seattle, WA
Bassetti & Morse
Hyperbolic paraboloid (barrel vault), multiple
    vaults with abutments
Demolished

1962   US Science Pavilion (World's Fair building)
200 Second Avenue N
Seattle, WA
Yamasaki & Associates
Precast concrete panels, cast-in-place slab

1963   Faber Hardware
Pasco, WA
Hyperbolic paraboloid, multiple

1963   Grant County Fairground
3953 Airway Drive NE
Moses Lake, WA
Multiple hyperbolic paraboloid (inverted
    umbrella)

1963   IBM Pittsburgh, Pedestrian Bridge
11 Boulevard of the Allies
Pittsburgh, PA
Curtis & Davis
Ribbed pedestrian bridge

1963 **Meadowdale Elementary School Classrooms**
6505 168th Street SW
Lynnwood, WA
Hyperbolic paraboloid (inverted umbrella)
Demolished

1963 **North Shore Congregation Israel**
1185 Sheridan Road
Glencoe, IL
Yamasaki & Associates
Fan-vaulted roof

1963 **Polaris Pacific Facility**
NBBJ
Hyperbolic paraboloid (inverted umbrella),
    single shells

1963 **Proctor Products Office**
210 Eighth Street
Kirkland, WA
Hyperbolic paraboloid (inverted umbrella),
    multiple

1963 **St. Frances Cabrini Church**
550 Paris Avenue
New Orleans, LA
Curtis & Davis
Cylindrical shell (modified), post-tensioned
    shells cantilevering 62 feet
Demolished

1964 **Christiansen Residence**
7797 Hansen Road NE
Bainbridge Island, WA
Hyperbolic paraboloid, roofs and floors

1964 **Cook Motors**
Mount Vernon, WA
Hyperbolic paraboloid, multiple

1964 **Furniture Store**
Everett, WA
Hyperbolic paraboloid, multiple

1964 **Olympic Junior College Gymnasium**
1600 Chester Avenue
Bremerton, WA
B & G Architects
Folded plate, multiple plates

1964 **Potato Storage Warehouse**
Moses Lake, WA
Hyperbolic paraboloid, multiple shells

1965 **Carleton College Gymnasium**
1 N College Street
Northfield, MN
Yamasaki & Associates
Groined vault (intersecting parabolic vaults),
    large shells

1965 **Pedestrian Bridge over Interstate 5**
NE 195th Street
Shoreline, WA
Double-cantilever construction method, hyper-
    bolic paraboloid underside

1966 **Geary Expressway Pedestrian Overpass**
1698 Geary Boulevard
San Francisco, CA
Yamasaki & Associates
Hyperbolic paraboloid bridge

1966 **Mercer Island Beach Club**
8326 Avalon Drive
Mercer Island, WA
Kirk, Wallace, McKinley & Associates
Hyperbolic paraboloid, clubhouse roofs and
    floors

1966 **Potato Processing Plant**
14124 Wheeler Road NE
Moses Lake, WA
Cylindrical (multiple), post-tensioned

1966 **Unique Frozen Foods**
811 W Gum Street
Connell, WA
Hyperbolic paraboloid, multiple shells,
    warehouse

1967 **International Exhibition Facility (Rivergate)**
900 Convention Center Boulevard
New Orleans, LA
Curtis & Davis
Cylindrical (multiple and continuous), post-
tensioned, exhibition hall
Largest-spanning long barrel in the world
Demolished

1967 **Tri-Cities Airport**
3601 N Twentieth Avenue
Pasco, WA
Pence & Stanley Associates
Hyperbolic paraboloid, multiple, terminal
building

1968 **First Federal Savings and Loan**
Kennewick, WA
Pence & Stanley Associates
Hyperbolic paraboloid, multiple

1968 **Freeway Motors**
Seattle, WA
Hyperbolic paraboloid, canopy
Demolished

1968 **Gloria Dei Lutheran Church**
6016 120th Avenue SE
Bellevue, WA
Kirk, Wallace, McKinley & Associates
Hyperbolic paraboloid, multiple

1968 **Ridgeway Lithograph**
4001 156th Avenue NE
Redmond, WA
Jacobsen
Hyperbolic paraboloid, multiple
Demolished

1969 **Chrysler Styling Center**
12800 Oakland Park Boulevard
Highland Park, MI
Yamasaki & Associates
Dome (segmental), double curved display
Demolished

1969 **Custom Manufacturing Building**
Seattle, WA
Jacobsen
Hyperbolic paraboloid, multiple

1969 **Glaser Beverages Distribution Center**
2300 Twenty-Sixth Avenue S
Seattle, WA
Sato
Hyperbolic paraboloid, multiple

1969 **Lamb Weston Inc.**
Seattle, WA
Hyperbolic paraboloid, multiple

1970 **Shopping Center**
Renton, WA
Hyperbolic paraboloid, multiple, covered mall

1970 **Ski Cabin**
Stevens Pass, WA
Hyperbolic paraboloid, floor structure,
residence

1971 **Lynnwood Industrial Center**
1300 N Ninety-Seventh Street
Seattle, WA
Hyperbolic paraboloid, multiple, warehouse

1971 **Nalley Valley Viaduct**
State Route 16
Tacoma, WA
Elevated roadway, tetrapod supports
Demolished

1972 **Nelco Manufacturing**
19220 Pioneer Way E
Orting, WA
Hyperbolic paraboloid, multiple

1972 **Reynolds Manufacturing**
5720 204th Street SW
Lynnwood, WA
Modified hyperbolic paraboloid, multiple,
manufacturing

1972   Temple Beth El
        7400 Telegraph Road
        Bloomfield Hills, MI
        Yamasaki & Associates
        Cable net, wood sheathed

1973   Formost Packaging Machines
        19211 144th Avenue NE
        Woodinville, WA
        Hyperbolic paraboloid, multiple

1973   H & N Hatchery
        Redmond, WA
        Hyperbolic paraboloid, multiple

1974   Highland Park High School
        Redmond, WA
        Cylindrical, multiple, school

1974   Transparent Bag Company
        11133 120th Avenue NE
        Kirkland, WA
        Hyperbolic paraboloid, multiple

1974   US Pavilion, Expo '74
        507 N Howard Street
        Spokane, WA
        NBBJ
        Cable net, fabric, fair pavilion

1976   King County Stadium (Kingdome)
        501 Stadium Place S
        Seattle, WA
        Naramore, Skilling & Praeger
        Dome (segmental), double curved
        Largest concrete dome in the world
        Demolished

1977   Sundstrand Data Control
        15001 NE Thirty-Sixth Street
        Redmond, WA
        Hyperbolic paraboloid, multiple

1979   Baltimore Convention Center
        1 W Pratt Street
        Baltimore, MD
        NBBJ
        Truncated pyramid, (three) halls,
            post-tensioned
        Altered

1985   Royal Saudi Naval Stadium
        Jubail, Saudi Arabia
        NBBJ
        Hyperbolic paraboloid, multiple,
            post-tensioned

1989   SunDome Arena
        1301 S Fair Avenue
        Yakima, WA
        LMN
        Dome, segmental

1991   Bainbridge Island High School Grandstand
        9332 High School Road NE
        Bainbridge Island, WA
        NBBJ
        Hyperbolic paraboloid, multiple, precast

2003   Chena River Bridge
        471 Wendell Avenue
        Fairbanks, AK
        Hyperbolic paraboloid bridge

NOTES

PREFACE

1 Rainer Metzger, "Jack Christiansen—Thin Shell Concrete in the Pacific Northwest," *Column 5* 20 (2006): 8–11.

2 Edward M. Segal, "The Thin Concrete Shells of Jack Christiansen" (master's thesis, Princeton University, 2008).

3 Edward M. Segal and David P. Billington, "Concrete Shell Roofs," in *Fifty Years of Progress for Shell and Spatial Structures: In Celebration of the 50th Anniversary Jubilee of the IASS (1959–2009)*, ed. Ihsan Mungan and John Fredrick Abel (Madrid: IASS, 2011), 154.

4 Christiansen, interview by Susan Boyle, June 15, 2009, "Modern Talk: Northwest Mid-Century Architects Oral History Project," Docomomo WEWA, Seattle.

5 Tyler S. Sprague, "A Preservation Survey of Hyperbolic Paraboloids in the Pacific Northwest," grant submitted to the Center for the Study of the Pacific Northwest, University of Washington, 2012.

6 *Structural Engineers of the 1962 Seattle World's Fair* (Seattle: Structural Engineers Foundation of Washington and ProMotion Arts, 2013), documentary film.

INTRODUCTION: UNCOVERING AN AMERICAN SHELL DESIGNER

1 Ralph Halpern, "Dome's Grand Opening Grand Showman," Magazine Section, *Seattle Times*, March 21, 1976.

2 Heather MacIntosh, "Kingdome: The Controversial Birth of a Seattle Icon (1959–1976)," *HistoryLink*, March 1, 2000, www.historylink.org/File/2164.

3 See John Chilton, *The Engineer's Contribution to Contemporary Architecture: Heinz Isler* (London: Thomas Telford, 2000).

4 John V. Christiansen, "Structural Form," in *Building Structural Design Handbook*, ed. Richard N. White and Charles G. Salmon (New York: John Wiley & Sons, 1987), 220.

5 The early history of thin shell concrete is an active research topic. For an outline of thin shell activities, see David P. Billington, *The Tower and the Bridge: The New Art of Structural Engineering* (New York: Basic Books, 1983). A more recent discussion can be found in Bernard Espion, "Pioneering Hypar Thin Shell Concrete Roofs in the 1930s," *Beton- und Stahlbetonbau* 111, no. 3 (2016): 159–65.

6 Recent scholarship includes Roland May, "Shell Sellers: The International Dissemination of the Zeiss-Dywidag System, 1923–1939," published pp. 557–64 in vol. 2 of the proceedings of the Fifth International Congress on Construction History,

Chicago, June 3–7, 2015. See also Eric M. Hines and David P. Billington, "Anton Tedesko and the Introduction of Thin Shell Concrete Roofs in the United States," *Journal of Structural Engineering* 130, no. 11 (2004): 1639–50.

7   Sigfried Giedion, "The State of Contemporary Architecture," *Architectural Record* 115, no. 2 (February 1954): 186–88.

8   See "Saarinen Challenges the Rectangle, Designs a Domed Auditorium and a Cylindrical Chapel for MIT's Laboratory Campus," *Architectural Forum* 98 (January 1953): 126–33, *Avery Index to Architectural Periodicals*, EBSCOhost. For further discussion, see "Three Critics Discuss M.I.T.'s New Buildings," *Architectural Forum* 104 (March 1956): 156, *Avery Index to Architectural Periodicals*, EBSCOhost. See also J. William Plunkett and Caitlin T. Mueller, "Thin Concrete Shells at MIT: Kresge Auditorium and the 1954 Conference," published pp. 127–36 in vol. 3 of the proceedings of the Fifth International Congress on Construction History, Chicago, June 3–7, 2015.

9   On the challenges and conversations surrounding the Sydney Opera House, see Anne Watson, ed., *Building a Masterpiece: The Sydney Opera House* (Sydney: Powerhouse Publishing in association with Lund Humphries, 2006).

10 Ada Louise Huxtable, "Concrete Technology in U.S.A.—Historical Survey," *Progressive Architecture* 41, no. 10 (October 1960): 144–49.

11 Stephen Buonopane and Mikhail Osanov, "Evaluation of August Komendant's Structural Design of the Shells of the Kimbell Art Museum," published pp. 283–90 in vol. 1 of the proceedings of the Fifth International Congress on Construction History, Chicago, June 3–7, 2015.

12 The Ingraham High School Gymnasium in Seattle contained post-tensioning strands within the shell thickness of cylindrical concrete shell vaults. Drawings of the structure are dated June 18, 1958.

13 David P. Billington, *Thin Shell Concrete Structures*, 2nd ed. (New York: McGraw-Hill, 1982), 341.

14 Deanna Duff, "Architecture at One Hundred," *Columns: The University of Washington Alumni Magazine*, September 2014.

15 *Modern Views: A Conversation on Northwest Modern Architecture* (Seattle: Studio/216, 2010), documentary film.

CHAPTER 1: BEGINNINGS

1   "Louise Linderoth, Christiansen Will Marry Tomorrow," *Daily Illini* (University of Illinois), June 25, 1926. At the time, student newspapers commonly published wedding announcements.

2   Jack Christiansen's grandfather, Soren Christiansen, had immigrated to Chicago from Denmark in 1889. Soren had grown up on the family farm, but as the second son, he was to receive only a small inheritance—spurring his move to the United States. Biographical information about Christiansen's family is taken from US Census records unless otherwise noted.

3   Later, in the 1950s and 1960s, Christiansen's father even took a more active role in the industry-wide marketing of dairy products. "C. Valdeman Christiansen" is listed as attending the "dairy marketing session" at the University of Illinois Agricultural Industries Forum on February 1–2, 1960, in *Technical Production of Milk*

Concentrates: Papers in Dairy Marketing (Urbana: Department of Agricultural Economics, College of Agriculture, University of Illinois, June 1961), 50.

4 Unassisted by modern medical treatments, Christiansen's father's heart condition ultimately led to his premature death at the age of sixty-three, in 1965.

5 The courses were funded by the James Nelson Raymond Lecture Fund, intended to encourage young artists. Christiansen's artwork was unfortunately not retained by the Art Institute or by the family.

6 The house had been built for Christiansen's Aunt Jo, who lived with Christiansen's paternal grandfather, Soren, until his passing. With more space than she needed, Aunt Jo moved to a smaller house in Oak Park so that Jack's family could move into the house on Woodbine.

7 The Christiansens' house was less than half a mile away from several Wright buildings, including the Frank Lloyd Wright Home and Studio, the Arthur B. Heurtley House, and the Walter Gale House. Despite this proximity, Christiansen did not recall any discussion of Wright's work at home before he attended the University of Illinois.

8 *Lost Horizon* was first published in 1933 and became available in paperback in 1939. The cover of the book Christiansen read likely showed a sketch of the dramatic mountain scene, with an airplane flying over, and a small city nestled within its valleys.

9 James Hilton, *Lost Horizon* (New York: William Morrow, 1933), 46–47.

10 "The Pacific Northwest: The Story of a Vision and a Promised Land," *Life* 6, no. 23 (June 5, 1939): 18, 20. While the article discussed the mountain ranges of the Northwest, including the Cascade Range, it actually depicted only the Pahsimeroi Mountains in Idaho, near the Salmon River, and the landscape surrounding the Grand Coulee Dam in eastern Washington.

11 Ibid., 18.

12 He worked in an egg-candling plant, where he inspected the quality of freshly laid eggs by shining a light through their shells. He worked in an M&M's/Mars candy factory—dumping kettles of melted chocolate into molds and then pushing the hot trays into the blast freezer to cool. His family never had any money problems at home, but young Christiansen appreciated the independence that came from earning his own wages.

13 "Newton Club," *Tabula—Oak Park and River Forest Township High School Yearbook*, vol. 51 (1945), 44.

14 Maynard Brichford, "A Brief History of the University of Illinois," 1970, University of Illinois Archives, http://archives.library.illinois.edu/features/history.php.

15 At the time, MIT, Columbia University, and Pennsylvania State University also had programs in architectural engineering.

16 Winton U. Solberg, *The University of Illinois, 1894–1904: The Shaping of the University* (Urbana: University of Illinois Press, 2000), 169.

17 M. S. Uihlein, "The University of Illinois, N. Clifford Ricker, and the Origins of Architectural Engineering Education in the United States," *Journal of Architectural Engineering* 22, no. 3 (2016), https://doi.org/10.1061/(ASCE)AE.1943-5568.0000175.

18 On the split between architects and engineers in France, see Antoine Picon, *French Architects and Engineers in the Age of Enlightenment*, Cambridge Studies in the

History of Architecture (Cambridge: Cambridge University Press, 1992). The French models of both architecture and engineering were highly influential in the United States.

19 Jill E. Pearlman, *Inventing American Modernism: Joseph Hudnut, Walter Gropius, and the Bauhaus Legacy at Harvard* (Charlottesville: University of Virginia Press, 2007), 2–3.

20 Pamela Lowry, "West Point and the Tradition of the Army Corps of Engineers," *21st Century Science and Technology*, Spring 2011, 40.

21 W. Bernard Carlson, "Academic Entrepreneurship and Engineering Education: Dugald C. Jackson and the MIT-GE Cooperative Engineering Course, 1907–1932," in *The Engineer in America*, ed. Terry S. Reynolds (Chicago: University of Chicago Press, 1991), 369.

22 Uihlein, "University of Illinois."

23 Anthony Alofsin, "Tempering the Ecole: Nathan Ricker at the University of Illinois, Langford Warren at Harvard, and Their Followers," in *The History of History in American Schools of Architecture, 1865–1975*, ed. Gwendolyn Wright and Janet Parks (New York: Temple Hoyne Buell Center for the Study of American Architecture and Princeton Architectural Press, 1990), 73–74. See also Roula Geraniotis, "The University of Illinois and German Architectural Education," *Journal of Architectural Education (1984–)* 38, no. 4 (1985): 15–21.

24 Alofsin, "Tempering the Ecole," 74.

25 Eric Mumford, "More than Mies: Architecture of Chicago Multifamily Housing, 1935–65," in *Chicago Architecture: Histories, Revisions, Alternatives*, ed. Charles Waldheim and Katerina Rüedi Ray (Chicago: University of Chicago Press, 2005), 82.

26 For discussion of such transition at the University of Washington, see Jeffrey Karl Ochsner, *Lionel H. Pries, Architect, Artist, Educator: From Arts and Crafts to Modern Architecture* (Seattle: University of Washington Press, 2007), 134.

27 Ricker led the Department of Architecture until 1910, but continued on as a professor. In the following years, the program was hotly debated between architecture and engineering departments, with the increasingly specialized field of structural engineering (distinct from civil engineering) challenging the field of architectural engineering. Yet the degree maintained its autonomy from the College of Engineering and was absorbed into the College of Fine and Applied Arts in the 1930s. Professors with prior professional experience, reinforcing the practical nature of the degree, continually led the program.

28 "Department and Curricula: Architecture," in *University of Illinois Annual Register, 1945–1946* (Urbana: University of Illinois, April 1946), 220–21.

29 Christiansen, interview by Boyle.

30 Christiansen could not recall the name of the instructor; Assistant Professors Mildred Holslag and Kathleen Post were both teaching at the time.

31 Author's correspondence with Christiansen.

32 Christiansen recalled that in both mechanical drawing and physics, during his first and second years, he was called to the front of the class and recognized for the highest score.

33 Christiansen particularly enjoyed the history classes. Professor Rexford Newcomb—a student of Ricker and second president of the Society of Architectural Historians—directed the history courses. Newcomb was a prolific scholar, writing

primarily on American regional topics. J. A. Chewning, "The Teaching of Architectural History during the Advent of Modernism, 1920s–1950s," *Studies in the History of Art* 35 (1990): 102, 109.

34 From the course description in "Architecture: Courses for Advanced Undergraduates and Graduates," in *University of Illinois Annual Register*, 276.

35 Christiansen makes this reference in the documentary film *Structural Engineers of the 1962 Seattle World's Fair* (2013).

36 Quoted in Kai-Uwe Bergmann, "Jack Christiansen: Elegance in Engineering," *Arcade* 20, no. 3 (2002): 34.

37 Quoted in David P. Billington, Hubert Shirley-Smith, and Philip N. Billington, "Bridge—Engineering," in *Encyclopaedia Britannica*, accessed September 17, 2018, www.britannica.com/technology/bridge-engineering.

38 Elizabeth Bauer Mock, *The Architecture of Bridges* (New York: Museum of Modern Art, 1949), 7.

39 Christiansen, interview by Boyle.

40 Frank Lloyd Wright, *Frank Lloyd Wright on Architecture: Selected Writings, 1894–1940*, ed. Frederick Guthheim (New York: Duell, Sloan & Pearce, 1941).

41 Henry-Russell Hitchcock, *In the Nature of Materials, 1887-1941: The Buildings of Frank Lloyd Wright* (New York: Duell, Sloan & Pearce, 1942).

42 Wright, *On Architecture*, 141.

43 "Curriculum in Architecture—Construction Option," in *University of Illinois Annual Register*, 223–24.

44 Christiansen also took Theory of Structures and Advanced Structures from Morgan. "Architecture: Courses for Advanced Undergraduates and Graduates," in *University of Illinois Annual Register*, 275–76.

45 "Alumni Directory," *University of Colorado Journal of Engineering* 16, no. 4 (September 1920): 17.

46 Leonard K. Eaton, *Hardy Cross: American Engineer* (Urbana: University of Illinois Press, 2006), vii.

47 Hardy Cross, *Engineers and Ivory Towers*, ed. Robert C. Goodpasture (New York: McGraw-Hill, 1952), 5.

48 Hardy Cross and Newlin Dolbey Morgan, *Continuous Frames of Reinforced Concrete* (New York: John Wiley & Sons; London: Chapman & Hall, 1932).

49 This method would have involved solving multiple equations all at once, using differential calculus. Ibid., 38.

50 Christiansen, interview by Boyle.

51 *The Illio*, vol. 52 (1945).

52 *Structural Engineers of the 1962 Seattle World's Fair* (2013), documentary film.

53 High honors signified a scholastic average of 4.25 according to the university.

54 Christiansen, interview by Boyle.

55 Later, in 1953, Baron would accept an invitation to move to the University of California, Berkeley, as a full professor and the director of the Structural Engineering Laboratory.

56 Christiansen's instructors included Professors Philip C. Rutledge (chairman), Lewis H. Kessler, and Frank Baron; Associate Professors Merrill B. Gamet, Richard S. Hartenberg, Wallis S. Hamilton, C. W. Muhlenbruch, and Jorj O. Osterberg; and Assistant Professors Howard P. Hall and William T. Hooper Jr.

57 Graduate course descriptions for academic year 1949–50 in *Northwestern University Bulletin* 49, no. 13 (December 27, 1948): 69–70.

58 Stephen Timoshenko, *Theory of Plates and Shells*, Engineering Societies Monographs (New York: McGraw-Hill, 1940).

59 See Thomas Leslie, *Chicago Skyscrapers, 1871–1934* (Urbana: University of Illinois Press, 2013).

60 "Perkins & Will Historical Outline," 2014, http://history.perkinswill.com/.

61 Scott Murray, *Contemporary Curtain Wall Architecture* (New York: Princeton Architectural Press, 2009), 43.

62 See May, "Shell Sellers."

63 Billington, *The Tower and the Bridge*, 173.

64 Hines and Billington, "Tedesko."

65 Maria Moreyra Garlock and David P. Billington, *Félix Candela: Engineer, Builder, Structural Artist* (Princeton, NJ: Princeton University Art Museum; New Haven, CT: Yale University Press, 2008), 30.

66 Hines and Billington, "Tedesko," 1642.

67 Anton Tedesko, "Shells 1970—History and Outlook," in *Concrete Thin Shells*, ed. Stefan J. Medwadowski, William C. Schnobrich, and Alex C. Scordelis (Detroit: American Concrete Institute, 1971), 3.

68 Tedesko designed US Air Force hangars in South Dakota (1948) and Maine (1948).

69 See Richard Weingardt, "Anton Tedesko: Father of Thin-Shell Concrete Construction in America," *Structure*, April 2007, 69–71, www.structuremag.org/article.aspx?articleID=439.

70 Charles S. Whitney, "Cost of Long-Span Concrete Shell Roofs," *Journal of the American Concrete Institute* (hereafter cited as *ACI Journal*), *Proceedings*, 46, no. 6 (June 1950): 765–76.

71 Félix Candela, "Simple Concrete Shell Structures," *ACI Journal* 48, no. 12 (December 1951): 321–31.

72 The couple had recently bought a car in Detroit, and rather than pay to have it shipped to the Northwest, Claus flew out to Michigan to drive it home himself.

73 Bergmann, "Christiansen," 34.

CHAPTER 2: FIRST FORMS

1 Ivarsson had just completed the engineering work on the south grandstand of the University of Washington stadium, with George Wellington Stoddard (1949).

2 See Tyler S. Sprague, "Products of Place: The Era of Reinforced-Concrete Skyscrapers in Seattle, 1921–1931," *Pacific Northwest Quarterly* 106, no. 3 (2015): 107–19.

3 For more on Skilling, see William Bain Jr., "John B. Skilling: Pioneering Building Engineer," in *Memorial Tributes: National Academy of Engineering* (Washington, DC: National Academy Press, 2010), 13:272–77, www.nap.edu/read/12734/chapter/48#277.

4 European activity in shell construction was clearly picking up at the time as well, and American engineers were exposed to that work, for example, through articles such as "Thin-Shell Concrete Roofs for French Roundhouses," *Engineering News-Record* 140, no. 21 (May 20, 1948): 752–53.

5 *Design of Cylindrical Concrete Shell Roofs*, Manuals of Engineering Practice, No.

31 (New York: American Society of Civil Engineers, 1952). The publication was prepared by the Committee on Masonry and Reinforced Concrete, within the Structural Division, through its Subcommittee on Thin Shell Design (which had only recently been formed).

6 The members of the subcommittee were distinguished engineers Charles S. Whitney, Chairman Hans H. Bleich, Alfred L. Parme, Mario G. Salvadori, and Herman Schorer. Christiansen maintained that the majority of writing and analytical work was carried out by Parme.

7 "History of Shell Design Theories," in *Design of Cylindrical Concrete Shell Roofs*, 3.

8 Not all concrete shells require edge beams, but ASCE Manual 31 adopted them as an integral part of the shell design process, likely due to the simplification of analysis procedure needed to make the manual practical for working design engineers.

9 Cawdrey & Vemo was the general contractor on several mid-scale projects in the Seattle area, including the Van Asselt Elementary School on Beacon Avenue and Orchard Street (1950), the First Methodist Church Parish House (1950), the King County Central Blood Bank (1950), the Pacific Telephone and Telegraph Warehouse (1952), the Seattle Art and Photo Supply Building (1953), and Seattle Nash Inc. (1953), and, in Renton, the St. Anthony Church Annex (1953).

10 "Swim Pool with Thin Shell Dome," *Pacific Architect and Builder* 61, no. 1 (January 1955): 18–19.

11 "Green Lake Pool Will Open Soon," *Seattle Times*, November 28, 1954.

12 "Going Up," *Seattle Times*, October 10, 1954.

13 "Swim Pool with Thin Shell Dome." The only other thin shell structure recorded in the area at that time was the Kenmore School, designed by architect Ralph Burkhard and engineer Peter H. Hostmark. Using a corrugated form (zigzags) of thin concrete, the spans were smaller, 92 feet 8 inches. "Concrete Slab Roof—Wavy Style," *Pacific Architect and Builder* 61, no. 12 (December 1955): 16–17. In the local newspapers, the only complaints stemmed from the interior acoustics. "Acoustics at Green Lake Pool Criticized," *Seattle Times*, December 3, 1954.

14 The firm's records make this claim in promotional material, though it is difficult to verify. If the claim of largest "intermediate vault" means that the proportions for an intermediate vault must be met (radius to length), then, presumably, this shell had the largest overall span. Tedesko had built barrel vaults of much longer span, though possibly not of "intermediate" proportions. ASCE Manual 31 was still quite new, so there were likely only a small number of designers practicing at the time.

15 For more on Maloney, see "Architect Biographies: John W. Maloney," Department of Archaeology and Historic Preservation, Washington State, https://dahp.wa.gov /historic-preservation/research-and-technical-preservation-guidance/architect -biographies/bio-for-john-w-maloney. A prolific designer, Maloney straddled many design eras, but his work is largely still unexplored.

16 The Shoreline School District's territory was annexed in 1954, resulting in the operation of eight additional schools (including two elementary schools, Viewlands and the new Olympic Hills), with three hundred new teachers and over nine thousand more students (forty-five hundred acquired directly through annexation).

17 Enrollment increased from twenty-two thousand students in 1922 to ninety thousand students by 1956.

18 "School Warehouse," *Seattle Times*, February 21, 1955.

19 The drawings clearly stated: "Details of formwork are suggestions only and contractor may use any other method he elects." This phrase legally allowed the contractor to build the formwork in whichever manner seen fit, though Christiansen knew his advice would be indispensable in ensuring an effective construction bid.

20 "New Public-School Warehouse Nears Completion," *Seattle Times*, February 5, 1956.

21 The *Seattle Times* published the cost for the district warehouse as $500,000. See ibid.

22 *Building Code of the City of Seattle* (Seattle: Seattle Building Department, 1943).

23 *Building Code of the City of Seattle* (Seattle: Seattle Building Department, 1956). This code shows a complete reorganization, addressing specifics of design and construction in a far more systematic (and modern) way than the preceding code.

24 Ibid.

25 Hostmark, who was born in Norway, moved to Seattle in 1927 and established his own firm, Peter H. Hostmark & Associates, in 1932. Marga Rose Hancock, "Century 21 World's Fair—Structural Engineering," *HistoryLink*, March 31, 2013, www.historylink.org/File/10363.

26 *Building Code of the City of Seattle* (1956), 173.

27 "New Building Code Approved," *Seattle Times*, September 11, 1956.

28 Anderson cofounded Concrete Technology Corporation in Tacoma, Washington, in 1951, along with his brother Thomas. A colleague of Belgian engineer Gustave Magnel (1889–1955), Anderson became an international expert in the technique and would create many signature structures, including the monorail beams for the 1962 Seattle World's Fair.

29 John V. Christiansen, interview with author, February 19, 2012.

30 While prestressing currently refers to tensioning cables before concrete is cast, and post-tensioning refers to tensioning after concrete is poured, there is less distinction made at this early point in the techniques' use. Despite its name, this conference paper arguably addresses the "post-tensioning" of Christiansen's vaults, as cables were embedded in a flexible conduit within the concrete edge beams and tensioned after concrete was poured. "ACI Regional Meeting in Seattle," *ACI Newsletter* 29, no. 3 (September 1957): 3–7. The abstract for Christiansen's presentation "Prestressing of Cylindrical Concrete Shell Structures" is published on page 7.

31 Anton Tedesko, "A Shell Review—Design and Construction Experiences," published pp. 99–110 in the proceedings of the Conference on Thin Concrete Shells, Cambridge, MA, June 21–23, 1954. See also "Prestressed Shell Roof at King's Lynn Factory," *Architects' Journal* 116 (August 7, 1952): 117.

32 T. Y. Lin, "Prestressed Concrete," *Scientific American* 199, no. 1 (1958): 25–31.

33 T. Y. Lin, "New Design Horizons through Prestressing of Concrete Shells," *Architect and Engineer* 218 (August 1959): 30–31. With Henry M. Layne, he also published "Prestressed Concrete Shell for Grandstand Roofs," *ACI Journal* 56, no. 11 (November 1959): 409–22.

34 "John Skilling: Architect's Engineer Plans the Biggest," *Engineering News-Record* 172, no. 14 (April 2, 1964): 124–25, 128.

35 "Northwest Engineers Turn to Thin-Shell Roofs," *Engineering News-Record* 156, no. 2 (January 12, 1956): 34–38.

36 "Undulated Thin-Shell Hangar Roof Spans Big Bays," *Engineering News-Record* 158, no. 10 (March 7, 1957): 44–46.

37 "New Ideas in Boeing Hangar Draw Interest," *Seattle Times*, August 5, 1956.

38 "Undulated Thin-Shell Hangar Roof."

39 Skilling publicized the hangar as the "largest concrete shell hangar in the world," though Tedesko had created barrel-vaulted hangars spanning 340 feet in Rapid City, South Dakota, and Limestone, Maine, in 1948.

40 A Vierendeel truss is distinct from a typical truss in that it has a rectangular, rather than triangular, geometry of members and openings—relying on rigid joints between vertical and horizontal members.

41 The church was noted in the *ACI Newsletter*, in "ACI Regional Meeting in Seattle," and was offered as a tour destination.

42 "Largest Single Thin Shell Concrete Design to Date," *Pacific Architect and Builder* 63, no. 2 (February 1957): 8.

43 The auditorium's roof is described in "Four-Hour Roof," *Seattle Times*, September 20, 1957. The reusable formwork is discussed in "How Jumbo Moves Thin Shell Forms," *Engineering News-Record* 158, no. 19 (May 9, 1957): 45.

CHAPTER 3: GEOMETRIC COMPLEXITY

1 Articles on thin shell concrete were published regularly in the *ACI Journal* from 1951 onward. These included George C. Ernst, "Stability of Thin-Shelled Structures," *ACI Journal* 49, no. 12 (December 1952): 277–91. In its February 1953 issue (vol. 49, no. 6), the *ACI Journal* published several articles addressing thin shells: Anton Tedesko, "Construction Aspects of Thin-Shell Structures" (505–20); Charles S. Whitney, "Reinforced Concrete Thin-Shell Structures" (521–36); Pier Luigi Nervi, "Precast Concrete Offers New Possibilities for Design of Shell Structures" (537–48); and Hermann Craemer, "Design of Prismatic Shells" (549–63).

2 Félix Candela, "Structural Applications of Hyperbolic Paraboloidical Shells," *ACI Journal* 51, no. 1 (January 1955): 397–416.

3 At the MIT conference—a seminal event in the dissemination of shell expertise in the United States—many architects, engineers, and builders from around the world (including Tedesko) presented their work. Candela delivered two papers: "The Shell as a Space Encloser" and "Discussion—Warped Surfaces: Hyperbolic Paraboloids," published pp. 5–11 and 91–98, respectively, in the proceedings of the Conference on Thin Concrete Shells, Cambridge, MA, June 21–23, 1954.

4 Garlock and Billington, *Candela*, 185–86.

5 The structural drawings were signed October 21, 1955. William Bain Sr. was the signing architect. The project team worked with Alvin Erickson as the local architect in Wenatchee.

6 Jeffrey Karl Ochsner, "Fred Bassetti," in *Shaping Seattle Architecture: A Historical Guide to the Architects*, ed. Jeffrey Karl Ochsner, 2nd ed. (Seattle: University of Washington Press, 2014), 334–39.

7 David A. Rash, "Morse, John Moore," in *Shaping Seattle Architecture*, 463–64.

8 Michael C. Houser, "Morse, John M. (1911–2000)," Docomomo WEWA, 2018, www.docomomo-wewa.org/architects_detail.php?id=46.

9 The firm started in residential home construction, gaining fame through two homes that were recognized by the American Institute of Architects with merit awards (the Marshall Forrest Residence in 1953 and the Gerald Martin Residence in 1954).

Bassetti & Morse had begun building on Mercer Island, a residential community that had been connected to Seattle in 1942 with the construction of the Interstate 90 floating bridge. In 1954, the institute's local chapter recognized the Lakeview School, located on Mercer Island, with a merit award—as designed by Bassetti, Morse, and an associate, William Aitken.

10 Bassetti was "closely associated with the development of Modern architecture in the Pacific Northwest from the 1940s to the 1990s." Ochsner, "Bassetti," 334.

11 Houser, "Morse."

12 The Lloyd Dodds Residence was recognized for its "exposed structure" as an "early example of the emerging Pacific Northwest regional interpretation of Modernism." Ochsner, "Bassetti," 335.

13 Model photograph and plan layout published in "February Start Planned," *Seattle Times*, December 1, 1957.

14 This project was the first that Worthington & Skilling had done with Bassetti & Morse. In addition to the thin shell work, the engineers at Worthington & Skilling were designing a large number of projects in the region, and it is likely that other architects recommended them.

15 "Structural Geometry for a High School Campus," *Pacific Architect and Builder* 65, no. 12 (December 1959): 27.

16 Ibid.

17 Ibid., 28.

18 Seven small buttresses around the perimeter supported the low point of each segment, and a below-grade ring tie resolved all outward thrust. Each four-sided segment connected to its neighboring segment through two sides, with the other two sides framing openings around the perimeter. The stiffening ribs ensured that both shell segments worked together.

19 "Structural Geometry for a High School Campus," 27.

20 Work by Dyckerhoff & Widmann in Germany used prestressing of shells in a similar manner. T. Y. Lin used a similar technique in the cylindrical vaults for the California Museum of Science and Industry, opened in 1961. T. Y. Lin and Felix Kulka, "Concrete Shells Prestressed for Load Balancing," in *World Conference on Shell Structures: Proceedings*, National Research Council Publication 1187 (Washington, DC: National Academy of Sciences–National Research Council, 1964), 423–30. See also Jürgen Joedicke, *Shell Architecture*, Dokumente der modernen Architektur 2 (New York: Alec Tiranti, 1963); Buonopane and Osanov, "Shells of the Kimbell Art Museum."

21 "Project Review: Auditorium," *Pacific Architect and Builder* 64, no. 3 (March 1958): 5.

22 "Plans OK'd for North Seattle High School," *Seattle Times*, June 7, 1958.

23 The sketch shows an auditorium with a slightly different configuration—four panels coming down onto four points, though one point is hidden from view. It is likely that other configurations were considered and that the sketch was simply an older one or not intended for detailed review. See ibid.

24 "Young Artists at Work," *Seattle Times*, November 11, 1959.

25 Nile Thompson and Carol J. Marr, "Edward S. Ingraham High School," in *Building for Learning: Seattle Public School Histories, 1862-2000* (Seattle: Seattle School District, 2002).

26 "Saddle-Shaped Shell," *Engineering News-Record* 157, no. 17 (October 25, 1956): 33; "Hyperbolic Paraboloid Roof Is Cheaper," *Engineering News-Record* 158, no. 15 (April 1957): 51; "Concrete Umbrella Roofs Test Builder's Ingenuity," *Construction Methods and Equipment* 39, no. 8 (August 1957): 86–88, 92.

27 Lawrence Lessing, "The Rise of Shells," *Architectural Forum* 109, no. 8 (August 1958): 106–11.

28 Anton Tedesko, "Shell at Denver—Hyperbolic Paraboloidal Structure of Wide Span," *ACI Journal, Proceedings,* 57, no. 4 (October 1960): 403–12.

29 Félix Candela, "Understanding the Hyperbolic Paraboloid," *Architectural Record* 123, no. 7 (July 1958): 191–95.

30 Ibid., 191.

31 For a history of what later became the International Association for Shell and Spatial Structures (and still known as the IASS), see Mungan and Abel, *Fifty Years of Progress for Shell and Spatial Structures.*

32 "Parking Planned," *Seattle Times,* November 9, 1957.

33 Worthington & Skilling, *University of Washington–City of Seattle Montlake Interchange: Preliminary Design* (Seattle: Worthington & Skilling, n.d. [1957]), 1, Special Collections, University of Washington Libraries.

34 The cantilevered sides of the bridge were similar to the cantilevered sections of T. Y. Lin's Rinconada Hippodrome in Caracas, Venezuela. Lin, "Prestressed Concrete."

35 "Mercer Island Residents Push Ambitious School Program," *Seattle Times,* September 8, 1963.

36 "$1,650,000 Project Set at Boeing Field," *Seattle Times,* May 31, 1960.

37 According to Christiansen: "A number of saddle surfaces may be fitted together along their edges . . . to form a vaulted roof of any length with span width. Such a structure is ideally configured to resist uniformly distributed load (such as snow) and nearly ideally configured to resist its self-weight. Any line in the surface parallel to the YZ plane is parabolic. Thus the only significant stresses are vault (Y direction) compressive stresses. Substantial resistance to unbalanced roof loads and lateral wind and seismic forces are available by virtue of the 'cylindrical shell'[–] like effective cross section. . . . The structure, consisting of a thin slab without arch ribs, has considerable flexural strength and stiffness. This vaulted structure has the potential to be very economical." John V. Christiansen, "Hyperbolic Paraboloid Thin Shell Concrete Structures for Sports Buildings," in *Space Structures for Sports Buildings: Proceedings of the International Colloquium on Space Structures for Sports Buildings, Oct. 27–30, 1987, Beijing, China,* ed. Tien T. Lan and Zhilian Yuan (Beijing: Science Press; London: Elsevier Applied Science, 1987), 187.

38 "Roof Shell Forming Details—Notes," Drawing Sheet S-8, *New Hangar and Shop Building for King County Airport,* Bassetti & Morse, September 11, 1961. The sheet is "checked" by "JVC," Christiansen's initials.

39 Christiansen, "Hyperbolic Paraboloid," 187.

40 Proctor was elected vice president of the Washington Aviation Association in 1951.

41 "Firm Building Kirkland Plant," *Seattle Times,* April 21, 1963.

42 An image of a model is shown in "Fall Start Scheduled," *Seattle Times,* June 7, 1959. *Pacific Architect and Builder* would keep its offices and printing operations in the building, documenting the construction within its own pages—in a series of articles published from December 1959 to January 1961.

43 Bergmann, "Christiansen," 34.

44 Kirk graduated from the University of Washington in 1937 and began collaborating with different architects in Seattle. He started his own firm as a sole practitioner in 1950. By 1957, the firm was known as Paul Hayden Kirk & Associates, and it eventually became Kirk, Wallace, McKinley & Associates, in 1960.

45 David A. Rash, "Paul Hayden Kirk," in Ochsner, *Shaping Seattle Architecture*, 296–300.

46 "Paul Kirk of Seattle," *Architectural Forum* 117, no. 2 (August 1962): 104–5.

47 C. L. Anderson, "'Parabs': Umbrellas of Concrete," *Seattle Times*, October 16, 1960.

48 Ibid.

49 "Paul Kirk of Seattle," 104–5.

50 Shell Forms Inc., "Shell Forms: Safe . . . Functional . . . Beautiful," advertising pamphlet, ca. 1980.

51 The Church of the Good Shepherd is a member of the General Convention of the Church of New Jerusalem, a liberal protestant sect nearly two hundred years old.

52 John V. Christiansen, "Shell Construction for the Church of the Good Shepherd," in *World Conference on Shell Structures*, 339–42.

53 The total cost was approximately $115,000.

54 "New Church Will Have Unusual Design," *Seattle Times*, February 15, 1961.

55 "Building Ideas Tested in Wood Research House," *Pacific Architect and Builder* 66, no. 12 (December 1960): 33; Margery Phillips, "Organized Living Space," *Seattle Times*, January 29, 1961.

56 "HP Roofs for Houses," *Progressive Architecture* 42, no. 8 (August 1961): 134–37.

57 "New Pool on Mercer Island," *Seattle Times*, July 24, 1966.

58 "New Talent: Engineers," *Architectural Forum* 115, no. 2 (August 1961): 92–94.

59 Ibid., 93.

60 Ibid.

CHAPTER 4: AN INTERNATIONAL SHOWCASE

1 Paula Becker, Alan J. Stein, and HistoryLink, *The Future Remembered: The 1962 Seattle World's Fair and Its Legacy* (Seattle: Seattle Center Foundation, 2011), 11–15.

2 Ibid., 15.

3 The slogan appeared on many different promotional materials for the fair, including brass tokens good for "one dollar in trade" on the fairgrounds.

4 *Century 21 Exposition* (Seattle: Century 21 Exposition Inc.; [Pasadena]: California Institute of Technology, 1959), 3.

5 Meredith Clausen, "Paul Thiry," in Ochsner, *Shaping Seattle Architecture*, 290–92.

6 "Seattle Votes for Architecture," *Architectural Record* 130 (August 1961): 98. The Design Standards Advisory Board consisted of four architects (John Detlie, Robert Dietz, Perry Johanson, and Paul Thiry); the City of Seattle planning director (John Spaeth); and two out-of-town design consultants, architect Minoru Yamasaki and landscape architect Lawrence Halprin.

7 *Century 21 Exposition*, 17.

8 For more on the landmark, see Knute Berger, *Space Needle: The Spirit of Seattle* (Seattle: Documentary Media, 2012).

9   Christiansen also worked with Bassetti & Morse on a dramatic, wood-shell entry arch, which was not completed.

10 *Century 21 Exposition*, 10.

11 For an in-depth look at Yamasaki's life and career, see Dale Allen Gyure, *Minoru Yamasaki: Humanist Architecture for a Modernist World* (New Haven, CT: Yale University Press, 2017).

12 Minoru Yamasaki, interview by Virginia Harriman, [ca. August 1959], transcript, Archives of American Art, Smithsonian Institution, www.aaa.si.edu/collections /interviews/minoru-yamasaki-interview-6235.

13 Yamasaki had earlier explored the medium in his design of the Dhahran Air Termi-nal in Saudi Arabia (1959–61). He had previously visited precast concrete facilities in Holland and was inspired to use the material in the Middle East—remarking on the excellent density and finish quality.

14 In 1956, Yamasaki was a guest speaker at a school-building conference in Yakima. "Ex-Seattleite Praises Idealism of State School Officials," *Seattle Times*, August 30, 1956.

15 "Civic Center Architects Well Qualified," *Seattle Times*, August 14, 1957.

16 "Hall of Science Architect Has Impressive Record," *Seattle Times*, January 23, 1960.

17 "Hall of Science Architect Named," *Seattle Times*, January 22, 1960.

18 Robertson had joined the firm only a few years earlier, but was quickly emerging as a dynamic and ambitious engineer who would take the reins of the World Trade Center project in New York City. See Leslie E. Robertson, *The Structure of Design: An Engineer's Extraordinary Life in Architecture* (New York: Monacelli Press, 2017).

19 "UW Alumnus of the Year Was 'Regular Fellow,'" *Seattle Times*, June 6, 1960.

20 Minoru Yamasaki, *A Life in Architecture* (New York: Weatherhill, 1979), 70.

21 "Architects 'Tickled to Death' with Their Science Pavilion," *Seattle Times*, April 5, 1962.

22 John L. Hutsell, "Fabrication of Science Pavilion Wall Panels," in "Concrete Con-struction for the Century 21 Exposition," ed. Harlan Edwards, *ACI Journal*, *Proceed-ings*, 60, no. 6 (June 1963): 686. This piece was published as part of a collection of seven brief papers highlighting concrete construction used at the 1962 Seattle World's Fair. All the papers were also presented at the fall 1962 convention of the ACI in Seattle.

23 Ibid.

24 Ibid., 692.

25 *Structural Engineers of the 1962 Seattle World's Fair* (2013), documentary film.

26 "Structure Plays Leading Role in Latest Yamasaki Designs," *Architectural Record* 134, no. 12 (December 1963): 103–10.

27 "Soaring Ribbed Vaults to Dominate Yamasaki's Design for Seattle Fair," *Architec-tural Record* 128, no. 8 (August 1960): 147–48.

28 Yamasaki, *A Life in Architecture*, 70.

29 Yamasaki, interview by Harriman.

30 Ibid.

31 Quoted in James Glanz and Eric Lipton, *City in the Sky: The Rise and Fall of the World Trade Center* (New York: Times Books, 2003), 120.

32 For more on Nervi's techniques, see Allison Halpern, David P. Billington, and Sigrid

Adriaenssens, "The Ribbed Floor Slab Systems of Pier Luigi Nervi," published in the proceedings of the International Association for Shell and Spatial Structures (IASS) Symposium 2013, Wrocław University of Technology, Poland, September 23–27, 2013; edited by J. B. Obrebski and R. Tarczewski.

33 "Architects 'Tickled to Death.'"

34 US Department of Commerce, "Architectural Critique of the Pavilion," in *United States Science Exhibit, Seattle World's Fair: Final Report* (Washington, DC: Government Printing Office, 1963), 51–56.

35 "Architects 'Tickled to Death.'"

36 Glanz and Lipton, *City in the Sky*, 88.

37 Bruce Morris Walker and John W. McGough, headed the firm. William Trogdon was a frequent collaborator during this time, but Trogdon is not recognized on the design documents for the World's Fair buildings. "Century 21 Contracts Approved," *Seattle Times*, February 26, 1960.

38 "Magazine Honors W.S.U. Dormitory," *Seattle Times*, July 17, 1960.

39 Maury Proctor, "Movable Forms for Six-Sided Hyperbolic Shells," in "Concrete Construction for the Century 21 Exposition," ed. Harlan Edwards, *ACI Journal, Proceedings*, 60, no. 6 (June 1963): 693–96.

40 Ibid., 693.

41 "An Architect's Guidebook to the Seattle World's Fair," special issue, *Architecture/West* (formerly *Pacific Builder and Engineer*) 66, no. 4 (April 1962): 32.

42 Proctor, "Movable Forms," 694.

43 Becker, Stein, and HistoryLink, *Future Remembered*, 248.

44 The statement continued: "With their classic Oriental overtones, [the shells] seem to be particularly fitting for the foreign displays they contain." "Architect's Guidebook to the Seattle World's Fair," 32.

45 "Art for 'Seattle Center' Campaign Under Way," *Seattle Times*, May 26, 1961.

46 "Paul Kirk of Seattle," 106–7.

47 Ibid., 103.

48 "3 New Projects for Century 21," *Progressive Architecture* 41, no. 12 (December 1960): 51.

49 Jack Christiansen, "A Post-tensioned Folded Plate Roof for the Seattle Civic Center," in "Concrete Construction for the Century 21 Exposition," ed. Harlan Edwards, *ACI Journal, Proceedings*, 60, no. 6 (June 1963): 706.

50 Ibid., 709.

51 "Concrete Construction for the Century 21 Exposition," Fifteenth Fall Meeting of the American Concrete Institute, Seattle, September 27–29, 1962.

52 Harlan Edwards, ed., introduction to "Concrete Construction for the Century 21 Exposition," *ACI Journal, Proceedings*, 60, no. 6 (June 1963): 673.

53 "Architects Get Honors for Work at Fair," *Seattle Times*, June 4, 1962.

CHAPTER 5: EXPANDING AUDIENCE

1 "Firm Changes Name to Include Partners," *Seattle Times*, April 6, 1967.

2 "Six New Yamasaki Projects," *Architectural Record* 130, no. 1 (July 1961): 126–40.

3 Yamasaki, *A Life in Architecture*, 81.

4 "A Synagogue by Yamasaki," *Architectural Record* 135, no. 3 (September 1964): 192.

5 "Men's Gymnasium Building, Carleton College, Northfield, Minn.," *Architectural Record* 137, no. 2 (February 1965): 129.

6 Ibid.

7 Ibid.

8 Charles K. Hyde, *Riding the Roller Coaster: A History of the Chrysler Corporation* (Detroit: Wayne State University Press, 2003), 211.

9 Ibid.

10 Charles Novacek, "Construction of the Walter P. Chrysler Building, Highland Park, Michigan," *ACI Journal* 69, no. 12 (December 1972): 777.

11 Ibid.

12 J. Richard Gruber, "Arthur Q. Davis and New Orleans: Advancing Regional Modernism," in *It Happened by Design: The Life and Work of Arthur Q. Davis*, by Arthur Q. Davis (Jackson: University Press of Mississippi in association with the Ogden Museum of Southern Art, University of New Orleans, 2009), xiii.

13 Davis, *It Happened by Design*, 89–93.

14 Nathaniel Curtis, *Nathaniel Curtis, FAIA: My Life in Modern Architecture; The Rivergate: Architecture and Politics, No Strangers in Pair-a-dice* (New Orleans: University of New Orleans; Tulane University, Howard-Tilton Memorial Library, 2002), 24.

15 This collaboration between architects is described in "Architectural Firm Opens Seattle Office," *Seattle Times*, July 28, 1963.

16 "A Contract," *Seattle Times*, July 31, 1961.

17 Davis, *It Happened by Design*, 26.

18 "New Prison at Shelton Boasts Latest in Security, Rehabilitation," *Seattle Times*, October 11, 1963.

19 This church was greatly damaged by Hurricane Katrina and demolished.

20 John V. Christiansen, "Shell Roof for International Exhibition Facility, New Orleans," in Medwadowski, Schnobrich, and Scordelis, *Concrete Thin Shells*, 139–40.

21 "Humpbacked Barrel Arches Cover 4-Acre Hall," *Engineering News-Record* 179, no. 2 (August 3, 1967): 58–60.

22 Leslie E. Robertson, "A Life in Structural Engineering," in *Seven Structural Engineers: The Felix Candela Lectures*, ed. Guy Nordenson (New York: Museum of Modern Art, 2008), 69.

23 Gruber, "Davis and New Orleans," xxi.

24 Ibid.

25 Davis, *It Happened by Design*, 47–48.

26 Gruber, "Davis and New Orleans," xxi.

27 John V. Christiansen, "Concrete Shell Roof for the International Exhibition Facility," presented at the Congreso Internacional sobre la Aplicacion de Estructuras Laminares en Arquitectura, IASS-IMCYC Memoria, Mexico City, September 1967.

28 "World's Largest Shell Roof," *Construction News*, December 14, 1967, 30–31.

29 "Humpbacked Barrel Arches," 59.

30 The Superdome was intended to make New Orleans a "major league city." The state legislature had approved $35 million for the Superdome, but Governor John McKeithen wanted a far larger, more elaborate stadium. Curtis & Davis completed its design and estimated the cost at $121 million, just for "bricks and mortar." McKeithen reapproached the state legislature and then approved the design. Gruber, "Davis and New Orleans," xxii.

31 Robert A. Barr, "Freeway Bridge Sets Pattern," *Seattle Times*, March 14, 1965.

32 "Link between Interstate 5 and Narrows Span Will Open Tomorrow," *Seattle Times*, October 21, 1971.

33 John V. Christiansen, "Nalley Valley Viaduct, Tacoma, WA," in "The Structural Designs of Jack Christiansen" (unpublished manuscript, 2010).

34 "Link between Interstate 5 and Narrows Span."

35 "Skilling, Helle, Christiansen, Robertson . . . an Exclusive BD&C Profile," *Building Design and Construction*, March 1970.

CHAPTER 6: THE GRANDEST SCALE

1 Seattle had maintained a professional baseball team, the Seattle Rainiers in the Pacific Coast League, since the 1920s; games were played at Sick's Stadium in Rainier Valley. However, in the 1950s and 1960s, the league was gradually eclipsed by Major League Baseball, and the Rainiers became a minor-league farm team for the Boston Red Sox.

2 The first national broadcast in color was the 1954 Tournament of Roses Parade, and the 1958 NFL championship was seen by approximately 45 million people nationwide.

3 Eric E. Duckstad and Bruce Waybur, *Feasibility of a Major League Sports Stadium for King County, Washington* (Menlo Park, CA: Stanford Research Institute, 1960), 3.

4 "$15,000,000 Bond Issue for Stadium Proposed to County," *Seattle Times*, October 15, 1959.

5 Ibid.

6 The actual construction costs of Candlestick are difficult to determine, with multiple figures published at the time. Yet $15 million was widely publicized. See "Candlestick Park," Ballparks of Baseball, accessed July 15, 2018, www.ballparksof baseball.com/ballparks/candlestick-park/.

7 Walt Crowley, *Rites of Passage: A Memoir of the Sixties in Seattle* (Seattle: University of Washington Press, 1995), 16.

8 "C.C. Group to Consider Bay Stadium," *Seattle Times*, March 5, 1963.

9 "Civic, Business Groups Join Ranks to Push Floating Arena," *Seattle Times*, March 5, 1963.

10 Ibid.

11 Quotations from speech by commission member John L. O'Brien given at sportswriters and sportscasters luncheon, January 20, 1964, Stadium File (1962-1963-1964), King County Municipal Archives.

12 For a full history of the Astrodome, see Robert C. Trumpbour and Kenneth Womack, *The Eighth Wonder of the World: The Life of Houston's Iconic Astrodome*, with a foreword by Mickey Herskowitz (Omaha: University of Nebraska Press, 2016).

13 *Astrodome* (Houston: Houston Sports Association, 1968), promotional material, copy in Joseph Gandy Collection, folder 22, Special Collections, University of Washington Libraries.

14 Louis O. Bass, "Unusual Dome Awaits Baseball Season in Houston," *Civil Engineering* 35, no. 1 (January 1965): 63–65. Bass was the ASCE's vice president and worked with Roof Structures Inc.

15 "2 Major Consulting Firms Due to Begin Stadium Study Here," *Seattle Times*, January 21, 1966.

16 Praeger, Kavanagh & Waterbury, "Engineering Study—Seattle–King County Stadium," March 31, 1966, Stadium File (1965-1966-1967), King County Municipal Archives.

17 "Domed-Stadium Cost around $30 million, Study Shows," *Seattle Times*, April 14, 1966.

18 Ibid.

19 "Another Way to Get a Stadium," *Seattle Times*, September 21, 1966.

20 On aluminum as an example, see John Peter and Paul Weidlinger, *Aluminum in Modern Architecture* (Louisville, KY: Reynolds Metals Company, 1956).

21 In 1961, the company began manufacturing and marketing architectural doors and paneling as part of an expanding product line and launched an architectural marketing program.

22 "Unique Concept in Wood: A Stadium Large Enough to Hold the Astrodome," *Weyerhaeuser Magazine* 65 (July 1967): 8.

23 "Weyerhaeuser to Design World's Largest Domed Stadium: Astrodome Would Fit Inside," *Seattle Times*, February 14, 1967.

24 John V. Christiansen, "Presentation to ASCE in Tacoma—May 2, 1967" (unpublished paper, 1967).

25 Ibid.

26 "Unique Concept in Wood," 9.

27 Christiansen, "Presentation to ASCE in Tacoma," 5.

28 "Huge Wood-Domed Arena Spanning 4 City Blocks Is Proposed," *New York Times*, March 5, 1967.

29 "Wood Dome Unveiled—but None for Sale," *Seattle Times*, April 11, 1967.

30 At 530 feet in diameter, the Tacoma Dome design was the largest free-spanning timber structure in the world. Its structural design, different from Christiansen's, uses a triangular, geodesic pattern of wood elements to make up the dome.

31 William H. Mullins, "The Persistence of Progressivism: James Ellis and the Forward Thrust Campaign, 1968–1970," *Pacific Northwest Quarterly* 105, no. 2 (2014): 55.

32 "Report of the Ad Hoc Committee for the Seattle Center Stadium Site," August 5, 1968, p. 3, Stadium File (1965-1966-1967), King County Municipal Archives.

33 "Report on Stadium Task Force," Central Association of Seattle, March 12, 1968, Stadium File (1965-1966-1967), King County Municipal Archives.

34 "King County Stadium: A Design Prospectus and Invitation to Submit Qualifications," King County Design Commission, November 1968, Stadium File (1965-1966-1967), King County Municipal Archives.

35 Sverdrup & Parcel had recently completed Busch Stadium in St. Louis (1964–66), with architect Edward D. Stone. Though uncovered, the stadium was built of reinforced concrete, with a "crown" of ninety-six thin shell concrete arches around the perimeter.

36 Praeger is not present in any design correspondence or publicity. In the later years, he was replaced in name by Richard Q. Praeger and John Waterbury, who took over as "joint venture partners." *King County Multipurpose Stadium* (King County Convention and Visitors Bureau), no. 7 (November 1974): 6. Author correspondence with Christiansen confirmed Praeger's lack of involvement.

37 King County Design Commission, press release, February 6, 1969, Stadium File (1965-1966-1967), King County Municipal Archives.

38 "Multipurpose Is Key to Stadium: Architect Sees Challenge," *Seattle Times*, March 4, 1969.

39 Ibid.

40 "Petition Filed Calling for Vote on Stadium Site," *Seattle Times*, August 16, 1969.

41 Naramore, Skilling & Praeger, *Schematic Design Phase I for the King County Multipurpose Stadium*, December 4, 1969, Stadium File (1965-1966-1967), King County Municipal Archives.

42 Ibid., 94.

43 Ibid., 96–97.

44 Ibid., 96.

45 "County Council Views Plans for Stadium," *Seattle Times*, December 3, 1969.

46 "2nd-Phase Stadium Contract Signed," *Seattle Times*, February 17, 1970.

47 "County Delays Stadium Action," *Seattle Times*, April 6, 1970.

48 Stadium Steering Committee, with George Briggs, vice president, "Minutes of Meeting—Tuesday, April 7, 1970," April 13, 1970, Stadium File (1965-1966-1967), King County Municipal Archives.

49 "Commission Approves Final Design for Domed Stadium," *Seattle Times*, May 15, 1970.

50 "Center Site for Domed Stadium Endorsed," *Seattle Times*, May 15, 1970.

51 Ibid.

52 Crowley, *Rites of Passage*, 44.

53 Construction cost index percentages extracted from published construction cost and building cost data in *Engineering News-Record*, 1959–68.

54 "Optimism Greater for Stadium Here," *Seattle Times*, February 1, 1971.

55 "County Council to Decide on Stadium Soon, O'Brien Predicts," *Seattle Times*, February 5, 1971.

56 "Stadium 1971," Stadium File (1965-1966-1967), King County Municipal Archives.

57 "Non-baseball King St. Study Set," *Seattle Times*, January 23, 1971.

58 John V. Christiansen, "The Kingdome," published in ASCE Fall Convention and Exhibit, San Francisco, October 17–21, 1977, preprint 2946, p. 2.

59 "Contract Signed for Designing Stadium," *Seattle Times*, December 10, 1971.

60 Christiansen, "Kingdome," 2.

61 Ibid.

62 "King St. Stadium May Get Thin Concrete Dome," *Seattle Times*, December 29, 1971.

63 "Largest Thin-Shell Dome Puts Lid on Stadium Cost Spiral," *Building Design and Construction*, June 1974, 62.

64 "Stadium May Get Thin Concrete Dome."

65 "Stadium-Design Documents Approved," *Seattle Times*, January 21, 1972.

66 Ibid.

67 "20 Firms Indicate Interest in Bid on Stadium Roof System," *Seattle Times*, February 16, 1972.

68 Christiansen, "Thin Shell Concrete Structures for Sports Buildings," 187.

69 Billington, *Thin Shell Concrete Structures*, 341.

70 Christiansen, "Kingdome," 15–16.

71 "How Large Will the Dome Be?," *Seattle Times*, July 16, 1972.

72 "King County Stadium," original drawing set, Stadium File (1965-1966-1967), King County Municipal Archives.

73 Christiansen, "Thin Shell Concrete Structures for Sports Buildings," 187.

74 "Design Team Employs 'Fast Track System,'" *King County Multipurpose Stadium* (King County Convention and Visitors Bureau), no. 7 (November 1974): 4.

75 William H. Mullins, *Becoming Big League: Seattle, the Pilots, and Stadium Politics* (Seattle: University of Washington Press, 2013), 252.

76 Christiansen, "Kingdome," 2–3.

77 *King County Multipurpose Stadium* (King County Convention and Visitors Bureau), no. 4 (July 1973): 4.

78 Ibid.

79 "Portland Firm's Stadium Bid Is Low," *Seattle Times*, August 15, 1972.

80 *King County Multipurpose Stadium* (King County Convention and Visitors Bureau), no. 4 (July 1973): 4. *Engineering News-Record* cost data reported a construction cost index of 947.56 in December 1964 and 1811.76 in December 1972.

81 This program was a product of the Department of Civil Engineering at MIT, under the direction of Charles L. Miller. Miller and his colleagues began an ambitious program to apply advancing computer capability to solving civil engineering problems. In 1964, the department initiated a product development group at MIT, called Integrated Civil Engineering Systems (ICES). *ICES STRUDL-1, the Structural Design Language* (Cambridge, MA: Department of Civil Engineering, Structures Division, MIT, 1967).

82 S. L. Chu, "Analysis and Design Capabilities of STRUDL Program," in *Numerical and Computer Methods in Structural Mechanics*, ed. Steven J. Fenves, Nicholas Perrone, Arthur R. Robinson, and William C. Schnobrich (New York: Academic Press, 1973), 229.

83 David E. Weisberg, "Civil Engineering Software Development at MIT," chap. 5 in *The Engineering Design Revolution: The People, Companies and Computer Systems That Changed Forever the Practice of Engineering* (self-published, 2008), www.cadhistory.net/.

84 Christiansen presented a case study of this procedure in 1972. John V. Christiansen, "The Structural Design of the APEC Office Building," presented at the Annual Meeting of Automated Procedures for Engineering Consultants, San Antonio, TX, November 15–17, 1972. See also Leslie E. Robertson, "Reflections on the World Trade Center," *Bridge* (National Academy of Engineering) 32, no. 1 (Spring 2002): 5–10.

85 ADL could be used as a "stand alone analysis and design system," or the results of STRUDL could be combined "with any ADL analysis for design by ADL." John V. Christiansen, Leslie E. Robertson, and Richard E. Taylor, "ADL: A Structural Analysis and Design System," published in ASCE National Structural Engineering Meeting, Cincinnati, OH, April 22–26, 1974, preprint 2244, pp. 1–19.

86 Ibid., 9. Their program consisted of both extensions to STRUDL and original design and analysis capabilities. These modifications adapted the input, output, and data storage capabilities of STRUDL to better suit the design project realities that consulting structural engineers faced. The program allowed for modification of core design criteria (material strengths, detailing requirements, etc.) at any point in the

design process, while retaining a cumulative database of the entire structure. This database made it easier to track the effects of certain changes (e.g., increased member size) and helped reduce computation time. It also provided output formatting (e.g., schedules of beam reinforcing and size tables) that could be directly included in contract documents.

87 Christiansen's computational work can be seen in John V. Christiansen, "Analysis of Lateral Load Resisting Elements," *PCI Journal* 18, no. 6 (1973): 54–71. In addition, each office had a remote job entry device, via telephone lines, which enabled Christiansen and others to use larger outside computers, as well as share their computing efforts between offices. By this time, Christiansen had written a library of sophisticated analysis programs, for both complex structures and routine structural design problems.

88 A large earthquake (measuring 6.5 on the Richter scale) struck the Puget Sound area in 1965, heightening awareness of the earthquake risks associated with large buildings like the Kingdome.

89 "Dome Design Beats Test," *Seattle Times*, May 13, 1973.

90 Ibid.

91 *King County Multipurpose Stadium* (King County Convention and Visitors Bureau), no. 1 (August 1972): 2.

92 *King County Multipurpose Stadium* (King County Convention and Visitors Bureau), no. 5 (February 1974): 5.

93 *King County Multipurpose Stadium* (King County Convention and Visitors Bureau), no. 6 (July 1974): 3.

94 Ibid., 1.

95 *King County Multipurpose Stadium* (King County Convention and Visitors Bureau), no. 7 (November 1974): 1.

96 "Stadium Dispute Heats Up over Design Responsibility," *Seattle Times*, December 5, 1974.

97 "Stadium: The House That Jacks Build," *Seattle Times*, January 5, 1975.

98 "Contractor May Finish Stadium—Spellman," *Seattle Times*, November 23, 1974.

99 "Stadium Work Gaining Steam," *Seattle Times*, December 18, 1974.

100 John C. Hughes, *John Spellman: Politics Never Broke His Heart* (Olympia: Washington State Heritage Center, Legacy Project, 2013), 149.

101 "Stadium Has a Name—the Kingdome," *Seattle Times*, June 27, 1975.

102 "The Sports Thing," *Seattle Times*, June 1, 1973.

103 "Stadium Dome Supports Itself," *Seattle Times*, July 22, 1975.

104 John Spellman, letter, *Kingdome Magazine* (King County Council), March 1976, 4.

105 Patrick Douglas, *Seattle Weekly*, Kingdome inaugural edition, March 31, 1976, quoted in "A Dome of Our Own," *Seattle Post-Intelligencer*, May 15, 1997.

106 Royal Brougham, "A Warm Welcome to the Big Dome," *Kingdome Magazine* (King County Council), March 1976, 6.

107 Billington, *Thin Shell Concrete Structures*, 342–43.

108 Ibid., 344.

109 David P. Billington, "Shells in Industry," *Bulletin of the International Association of Shell and Spatial Structures* 22, no. 7 (August 1979): 14–15.

110 *King County Multipurpose Stadium* (King County Convention and Visitors Bureau), no. 8 (June 1975): 1.

111 Bergmann, "Christiansen," 34.

112 John V. Christiansen, "The King County Stadium (Kingdome)," in Christiansen, "Structural Designs of Jack Christiansen."

CHAPTER 7: RETIREMENT AND PERSEVERANCE

1 Alfred L. Parme, "International Colloquium on Progress of Shell Structures in the Last 10 Years and Its Future Development," General Report, Session III (Canada, United States), Madrid, 1969, *Bulletin of the International Association for Shell and Spatial Structures*, no. 47 (December 1971): 49–51.

2 Tedesko, "Shells 1970," 3.

3 Ibid., 8.

4 Ibid., 13.

5 Other presenters included Félix Candela, David Billington, Ray W. Clough, and Alex C. Scordelis.

6 A discussion of this name change is included in Ekkehard Ramm, "The Decade 1970–1979: From 'Shell' to 'Shell and Spatial' Structures," in Mungan and Abel, *Fifty Years of Progress for Shell and Spatial Structures*, 40.

7 The US Pavilion was designed by architects Davis, Brody, Chermayeff, Geismar, deHarak Associates of New York City, with structural engineers David Geiger and Horst Berger.

8 William Addis, *Building: 3000 Years of Design Engineering and Construction* (London: Phaidon, 2007), 597.

9 See William J. R. Curtis, "Crises and Critiques in the 1960s," chap. 26 in *Modern Architecture since 1900*, 3rd ed. (London: Phaidon, 1996).

10 "Industrial Property—Pre-engineered-Fireproof," Shell Forms Inc., classified listing, *Seattle Times*, November 28, 1972.

11 Advertisement, *Seattle Times*, January 21, 1973. In April, Proctor changed the advertisement to promise spans to "100 feet." Advertisement, *Seattle Times*, April 5, 1973.

12 The advertisements ran on September 16, 20, 25, and 30, 1973.

13 Advertisement, *Seattle Times*, December 16, 1973.

14 The title line of the advertisement was changed to "New Commercial Bldg?" and later, in April 1973, to simply "LOW-COST All-Concrete Indust. Bldgs." The final listing ran in the *Seattle Times* on May 7, 1974.

15 Christiansen modified the typical umbrella design for the Reynolds Manufacturing warehouse, built in Lynnwood, Washington, in 1972. This warehouse was profiled in Polly Lane, "Umbrella Opens Wider," *Seattle Times*, November 19, 1972.

16 Figure published in John V. Christiansen, "Hyperbolic Paraboloid Performance and Cost," in *Hyperbolic Paraboloid Shells: State of the Art*, ed. John V. Christiansen (Detroit: American Concrete Institute, 1988), 144.

17 Proctor Products listed a job opening for "journeyman welder and metal fabricator," on March 16–20, 1973, and then for "metal fabrication foreman," on February 7–20, 1974. Job openings for "metal fabricators" were listed throughout 1978.

18 Yamasaki, *A Life in Architecture*, 144.

19 "Expo '74: Nature Festival," *Progressive Architecture* 55, no. 8 (August 1974): 74–[77].

20 US Department of Commerce, *Final Report from the Secretary of Commerce to the Congress on the United States Pavilion, Expo '74, International Exposition, Spokane, Washington, May 4–November 3, 1974* (Washington, DC: Government Printing Office, 1975).

21 William Bain Jr., John V. Christiansen, and Arthur K. Fisher, "The United States Pavilion," *Consulting Engineer* 43, no. 1 (July 1974): 65–68.

22 The Department of Commerce initially selected the Portland office of Skidmore, Owings & Merrill to design the US Pavilion. US Department of Commerce, *Final Report*, 6.

23 "US Pavilion at Expo '74," project description, n.d., Jack Christiansen files.

24 "Expo '74 Rejuvenates Downtown Spokane," *Civil Engineering* 44, no. 6 (June 1974): 48–50.

25 "Expo '74 Site Near Completion," *Seattle Times*, April 14, 1974.

26 US Department of Commerce, *Final Report*, 14.

27 W. J. Granberg, "His Buildings Have an Element of Surprise," *Seattle Post-Intelligencer*, July 7, 1974.

28 Author's correspondence with Christiansen.

29 Nelsen was a prominent Seattle architect. Born in Ruskin, Nebraska, he received his training in architecture at the University of Oregon and moved to Seattle soon after graduating in 1951. By 1953, Nelsen had established his own architectural practice in the University District. Over the course of the next four decades, he worked independently or in partnership with others to produce a number of notable works.

30 Polly Lane, "The Great Gallery—$20 Million Structure Will House Museum of Flight," *Seattle Times*, July 2, 1985.

31 Polly Lane, "An Imaginative Mix of Old, New," *Seattle Times*, December 9, 1979.

32 For example, Joseph Madl Jr., space frame structure, US Patent 4070847A, filed in 1976, published January 31, 1978.

33 Vladimir Bazjanac, "Daylighting Design for the Pacific Museum of Flight, Energy Impacts" (Berkeley: Applied Science Division, Lawrence Berkeley Laboratory, University of California, September 1988).

34 Herb Robinson, "Museum of Flight—Steel, Glass and a Hint of Banana Oil," *Seattle Times*, February 23, 1986.

35 Bill Dietrich, "A Museum Takes Flight—Stunning New Great Gallery on Boeing Field Is a Soar Spot the Region Will Be Proud to Show Off," *Seattle Times*, July 8, 1987.

36 John V. Christiansen, "25 Years of Shell Structures" (seminar, School of Engineering, Princeton University, Princeton, NJ, April 1979).

37 John V. Christiansen, "A Comparative Study of Large Covered Stadia" (seminar, School of Engineering, Department of Architecture, Princeton University, Princeton, NJ, April 1979).

38 "Royal Saudi Naval Stadium," project description, n.d., Christiansen Personal Archive.

39 Billington, *Thin Shell Concrete Structures*, xiii.

40 *National Geographic* 159, no. 1 (January 1981).

41 In addition to teaching structural design classes, Christiansen taught a course in building materials and assembly processes, from January to March 1986.

42 Christiansen presented "Design and Construction of the Kingdome," as a civil engineering seminar at Purdue University in March 1984. He presented "Preliminary

Structural Design Techniques—Structural Form" at the University of Wisconsin–Madison in November 1985.

43 Christiansen, "Thin Shell Concrete Structures for Sports Buildings," 186–93.

44 Christiansen, *Hyperbolic Paraboloid Shells*.

45 Ibid., iii.

46 John V. Christiansen, "Economics of Hyperbolic Paraboloid Concrete Shells," *Concrete International* 12, no. 8 (1990): 25.

47 "Thin-Shell Concrete Dome Built Economically with Rotating Forming and Shoring System," *Concrete Construction*, June 1991, 490–92.

48 Ibid.

49 Christiansen cited David H. Geiger, "A Cost Comparison of Roof Systems for Sports Halls," in Tan and Yuan, *Space Structures for Sports Buildings*, 492.

50 "High School Grandstand Is Cost Effective," Architecture-Engineering Section, *Daily Journal of Commerce*, November 27, 1991.

51 Bergmann, "Christiansen," 34.

52 Christiansen, "Economics of Hyperbolic Paraboloid Concrete Shells," 25.

53 Shelby Scates, "Engineer Climbing High, Moves from Skyscrapers to Olympics," *Seattle Post-Intelligencer*, October 8, 1989.

54 Suzanne Downing, "Local Engineer Reaches a New Peak with 100th Olympics Climb," *Bainbridge Island Review*, October 25, 1989.

55 John V. Christiansen, "Aesthetic Choices Leading to the Kingdome," published in ASCE Spring Convention, New York, May 11–15, 1981, preprint 81-072, pp. 1–9.

56 Ibid., 1. In the article, Christiansen becomes emphatically defensive of the Kingdome as a work of both architecture and engineering—drawing from engineering logics and principles of artistic composition in attempt to make his case.

57 Ray Faulkner and Edwin Ziegfeld, *Art Today: An Introduction to the Visual Arts*, 5th ed. (New York: Holt, Rinehart and Winston, 1969), 278.

58 Christiansen, "Aesthetic Choices Leading to the Kingdome," 9.

59 "Longspans: Kingdome's 'Thin Shell' Designed like an Egg," *Seattle Times Pacific Magazine*, April 25, 1982, 34.

60 To make matters worse, on August 19, two workers removing the ceiling tiles fell to their deaths when a crane cable broke—in an issue unrelated to the structural performance of the concrete dome.

61 Timothy Egan, "A Dome for All Seasons, but Not for All Time," *New York Times*, August 26, 1994.

62 Michael Janofsky, "An Open-and-Shut Showcase: Toronto's New SkyDome Is Built to Weather the Elements," *New York Times*, June 2, 1989.

63 Rene Motro, "The Decade 2000–2009: The Start of a New Century," chap. 6 in Mungan and Abel, *Fifty Years of Progress for Shell and Spatial Structures*, 113.

64 Washington State Major League Baseball Stadium Public Facilities District (Seattle), "Finance," accessed September 13, 2018, www.ballpark.org/finance.aspx.

65 *Engineering News-Record* cost data reported a construction cost index of 2494.3 in 1976 and 5999.63 in 1999.

66 Barbara Kingsley, "Is It a Dome Fit for King County? NFL Officials Say Yes, but Owner Behring, Seeking to Get out of His Lease, Claims the Kingdome Is Unsafe for His Team and Fans," *Orange County Register*, March 11, 1996.

67 Engineers at BergerABAM (Bob Mast), RSP/EQE (John Hooper), and Shannon &

Wilson issued a final report in July 1995. "Seattle's Kingdome Receives Rigorous Seismic Study," *Civil Engineering* 66, no. 4 (1996): 12.

68 Clare Farnsworth, "County Makes Its Case to the NFL," *Seattle Post-Intelligencer*, March 14, 1996.

69 "Seismic Study Urges Work on Kingdome—Recommendations 'Mild to Moderate' in Nature," *Seattle Post-Intelligencer*, July 26, 1995.

70 "Built to Last," *Seattle Times*, March 19, 2000.

71 "Designers: Why Dump an Engineering Marvel? The Man Who Crafted the Kingdome Says the Structure Has Many Good Years Left," *Seattle Times*, March 30, 1997.

72 "Backers Come to the Defense of the Dome," *Seattle Post-Intelligencer*, April 14, 1997.

73 "Our Readers' Views," *Columbian* (Vancouver), June 3, 1997.

74 "Dome's Still Big League to Him," *Seattle Times*, March 25, 1998.

75 "Landmark's Rise and Fall," *Oregonian*, March 23, 2000.

76 "Dome's Still Big League to Him."

77 "Dome Saver," *Seattle Post-Intelligencer*, May 27, 1997.

78 "Dome's Still Big League to Him."

79 "Dome Saver."

80 "Dome's Still Big League to Him."

81 "Kingdome Stories," *Seattle Times Pacific Magazine*, May 18, 1997.

82 "Down with the Dome," *Seattle Times*, March 24, 2000.

83 Sam Howe Verhovek, "Doomed Domes of the 'Old' Diamonds: Build It and They May Come," *New York Times*, July 13, 1999.

84 "Landmark's Rise and Fall."

85 "For Designer of Kingdome, Its Demise Blows Him Away," *Seattle Times*, March 31, 2000.

86 Ibid.

87 "Dome's Final Roar," *Seattle Times*, March 27, 2000.

88 "Landmark's Rise and Fall."

CONCLUSION: THE LEGACY AND FUTURE OF CONCRETE SHELLS

1 Christiansen worked with Alaska-based PDC Inc. Engineers and the artist Ron Senungetuk.

2 "Kingdome Considered Icon by Some Engineers," *Seattle Times*, August 6, 1998.

3 Ibid.

4 Segal and Billington, "Concrete Shell Roofs," 157, 162.

5 R. Sundaram, "Concrete Shell Roofs in Asia," in *Proceedings of the International Association of Shell and Spatial Structures (IASS) Symposium: Spatial Structures—Temporary and Permanent*, ed. Q. Zhang, L. Yang, and Y. Hu (Beijing: China Architecture & Building Press, 2010), 116–30.

6 Christian Meyer and Michael Sheer, "Do Concrete Shells Deserve Another Look?," *Concrete International*, October 2005, 43–50.

7 Gabriel Tang, "An Overview of Historical and Contemporary Concrete Shells, Their Construction and Factors in Their General Disappearance," *International Journal of Space Structures* 30, no. 1 (2015): 1–12.

8 "Meiso no Mori Municipal Funeral Hall, Toyo Ito and Associates, Architects," *Archi-*

tect Magazine, March 17, 2013, www.architectmagazine.com/project-gallery /meiso-no-mori-municipal-funeral-hall.

9 Edward Allen, Wacław Zalewski, and Boston Structures Group, *Form and Forces: Designing Efficient, Expressive Structures* (Hoboken, NJ: John Wiley & Sons, 2010).

10 Sigrid Adriaenssens, Philippe Block, Diederik Veenendaal, and Chris Williams, eds., *Shell Structures for Architecture: Form Finding and Optimization* (London: Routledge, 2014).

11 Other American recipients include the academics Stefan J. Medwadowski (1991) and Alex C. Scordelis (1994), the designer of tensile structures Horst Berger (2012), and professor, researcher, and author John Abel (2013).

12 Nadine Post, "Jack Christiansen, Leader in Thin Concrete Shells, Dead at 89," *Engineering News-Record*, August 30, 2017, www.enr.com/articles/42655/; "Jack Christiansen's Designs Were 'Grand Scale' Sculpture," Architecture and Engineering, *Daily Journal of Commerce*, August 31, 2017, www.djc.com/news/ae/12103705 .html; Mike Lindblom, "Pioneering Engineer Who Designed Kingdome, Museum of Flight Roofs, Has Died," *Seattle Times*, September 2, 2017, www.seattletimes.com/; "Celebrated Designer of Seattle Kingdome Dies at 89," *ASCE News*, September 14, 2017; http://news.asce.org/.

13 Post, "Christiansen."

14 Leslie E. Robertson and Jon Magnusson, correspondence with author.

ARCHIVAL AND BIBLIOGRAPHIC SOURCES

CHRISTIANSEN PERSONAL ARCHIVE

Includes a wide variety of personal and professional material collected from Christiansen's home office: a textbook and notes from Christiansen's education, early family photos, clipped articles from professional journals dating from the early 1950s, project description sheets; from his time at the Skilling firm (1954–82), firm-produced publicity material, journal articles, subject files, and slides of projects under construction; post-retirement, project drawings for Yakima SunDome (1989), Bainbridge Island High School Grandstand (1991), and Chena River Bridge (2003); also speculative projects, Christiansen's continued work with analytical software packages, slides of projects under construction, and more than one hundred slide carousels documenting Christiansen's climbing and international travel (dating from 1955).

KING COUNTY ARCHIVES, SEATTLE

Documentation, correspondence, reports, photographs, and architectural drawings associated with the Kingdome, primarily in the following Department of Stadium Administration series: Series 473, Public Information; Series 497, Kingdome Architectural Files and Drawings; Series 1608, Tenant Services Division.

PAUL HAYDEN KIRK PAPERS, SPECIAL COLLECTIONS, UNIVERSITY OF WASHINGTON LIBRARIES, SEATTLE

Documents significant to Christiansen and Kirk joint projects, including the Church of the Good Shepherd, United Control Corporation, Wood Products Research House, and Fine Arts Pavilion for the Seattle World's Fair.

MKA ARCHIVES, SEATTLE

Christiansen's former firm, still in practice as Magnusson Klemencic Associates (MKA). Records held include extensive photo and slide library (MKA Slide Archive), which documents Christiansen's early shell work; microfilmed drawings; and publicity material, historical publications, and limited correspondence.

SEATTLE PUBLIC SCHOOL ARCHIVE, SEATTLE

Includes construction drawings and photographs of several Christiansen projects (under construction and completed): Seattle School District Warehouse, Chief Sealth High School, Ingraham High School, and others.

AUTHORED OR CO-AUTHORED BY CHRISTIANSEN

Christiansen, J. V., and J. B. Skilling. Abstract for "Prestressing of Cylindrical Concrete Shell Structures." In "ACI Regional Meeting in Seattle," *ACI Newsletter* 29, no. 3 (September 1957): 3–7.

Christiansen, J. V., and J. B. Skilling. "A Concrete Shell Hangar for Jet Aircraft." Presented at the Building Research Institute Conference on Construction of Thin Shell Concrete Structures, Washington, DC, April 1963.

Christiansen, J. V. "A Post-tensioned Folded Plate Roof for the Seattle Civic Center." In "Concrete Construction for the Century 21 Exposition," edited by Harlan Edwards. *Journal of the American Concrete Institute, Proceedings*, 60, no. 6 (June 1963): 705–9.

Christiansen, J. V. "Shell Construction for the Church of the Good Shepherd." In *World Conference on Shell Structures: Proceedings*, 339–42. National Research Council Publication 1187. Washington, DC: National Academy of Sciences–National Research Council, 1964.

Christiansen, J. V. "Concrete Shell Roof for the International Exhibition Facility." Presented at the Congreso Internacional sobre la Aplicacion de Estructuras Laminares en Arquitectura, IASS-IMCYC Memoria, Mexico City, September 1967.

Christiansen, J. V. "Shell Roof for International Exhibition Facility, New Orleans." In *Concrete Thin Shells*, edited by Stefan J. Medwadowski, William C. Schnobrich, and Alex C. Scordelis, 139–48. Detroit: American Concrete Institute, 1971.

Christiansen, J. V. "Cast in Place, Reinforced Concrete Systems." In *State of the Art Report, Committee 3*, produced for the ASCE-IABSE International Conference on Planning and Design of Tall Buildings, Lehigh University, Bethlehem, PA, August 21–26, 1972.

Christiansen, J. V. "The Structural Design of the APEC Office Building." Presented at the Annual Meeting of Automated Procedures for Engineering Consultants, San Antonio, TX, November 15–17, 1972.

Christiansen, J. V. "Analysis of Lateral Load Resisting Elements." *PCI Journal* 18, no. 6 (1973): 54–71.

Christiansen, J. V., L. E. Robertson, and R. E. Taylor. "ADL: A Structural Analysis and Design System." Published in ASCE National Structural Engineering Meeting, Cincinnati, OH, April 22–26, 1974. Preprint 2244, pp. 1–19.

Christiansen, J. V., and L. E. Robertson. "Design for Wind Loading in Tall Buildings." Published in the proceedings of the Joint WSCSEA and NWCSEA Western Roundup, Sun River, Oregon, June 1974.

Christiansen, J. V., W. Bain Jr., and A. K. Fisher. "The United States Pavilion: Expo '74." *Consulting Engineer* 43, no. 1 (July 1974): 65–68.

Christiansen, J. V. "Seismic Analysis and Design of Concrete Building Structures." Presented at Seminario IMCYC-ACI, Mexico City, December 1974.

Christiansen, J. V. "King County Multipurpose Domed Stadium." Published pp. 1049–61 in vol. 2 of the proceedings of the IASS (International Association of Shell and Space Structures) World Congress on Space Enclosures (WCOSE-76), Montreal, July 1976.

Christiansen, J. V. "The Kingdome." Published in ASCE Fall Convention, San Francisco, October 1977. Preprint 2946, pp. 1–26.

Christiansen, J. V., Fritz Reinitzhuber, and Walter P. Moore Jr. "Structural Systems." *Infotech State of the Art Report*, 1980, 1–61.

Christiansen, J. V. "Aesthetic Choices Leading to the Kingdome." Published in ASCE Spring Convention, New York, May 1981. Preprint 81-072, pp. 1–11.

Christiansen, J. V. "Hyperbolic Paraboloid Thin Shell Concrete Structures for Sports Buildings." In *Space Structures for Sports Buildings: Proceedings of the International Colloquium on Space Structures for Sports Buildings, Oct. 27–30, 1987, Beijing, China*, edited by Tien T. Lan and Zhilian Yuan, 186–93. Beijing: Science Press; London: Elsevier Applied Science, 1987.

Christiansen, J. V. "Structural Form." In *Building Structural Design Handbook*, edited by Richard N. White and Charles G. Salmon, 216–36. New York: John Wiley & Sons, 1987.

Christiansen, J. V. "Hyperbolic Paraboloid Performance and Cost." In *Hyperbolic Paraboloid Shells: State of the Art*, edited by J. V. Christiansen, 139–58. Detroit: American Concrete Institute, 1988.

Christiansen, J. V. "Economics of Hyperbolic Paraboloid Concrete Shells." *Concrete International* 12, no. 8 (1990): 24–29.

DISCUSSIONS OF CHRISTIANSEN'S WORK

Houser, Michael. "Christiansen, John (1927–2017)." Docomomo WEWA, 2018. www.docomomo-wewa.org/architects_detail.php?id=53.

Melaragno, Michele G. "John V. Christiansen." In *An Introduction to Shell Structures: The Art and Science of Vaulting*, 192. New York: Van Nostrand Reinhold, 1991.

Metzger, Rainer. "Jack Christiansen—Thin Shell Concrete in the Pacific Northwest." *Column 5* 20 (2006): 8–11.

Segal, Edward M. "The Thin Concrete Shells of Jack Christiansen." Master's thesis, Princeton University, 2008.

Segal, Edward M. "The Thin Concrete Shells of Jack Christiansen." Published in the proceedings of the International Association for Shell and Spatial Structures (IASS) Symposium, September 28–October 2, 2009, Valencia, Spain.

Segal, Edward M., and David P. Billington. "Concrete Shell Roofs." In *Fifty Years of Progress for Shell and Spatial Structures: In Celebration of the 50th Anniversary Jubilee of the IASS (1959–2009)*, edited by Ihsan Mungan, and John Fredrick Abel, 154–64. Madrid: IASS, 2011.

Sprague, Tyler S. "'Beauty, Versatility, Practicality': The Rise of Hyperbolic Paraboloids in Post-war America (1950–1962)." *Construction History* 28, no. 1 (2013): 165–84.

Structural Engineers of the 1962 Seattle World's Fair. Documentary film. Seattle: Structural Engineers Foundation of Washington and ProMotion Arts, 2013.

THIN SHELL CONCRETE

Billington, David P. *Thin Shell Concrete Structures*. 2nd ed. New York: McGraw-Hill, 1982. First edition published in 1965.

Billington, David P., and Maria Moreyra Garlock. "Thin Shell Concrete Structures: The Master Builders." *Journal of the International Association for Shell and Spatial Structures* 45, no. 146 (2004): 147–55.

Boothby, Thomas E., M. Kevin Parfitt, and Mark Ketchum. "Milo S. Ketchum and Thin-Shell Concrete Structures in Colorado." *APT Bulletin* 43, no. 1 (2012): 39–46.

Chilton, John. *The Engineer's Contribution to Contemporary Architecture: Heinz Isler.* London: Thomas Telford, 2000.

Espion, Bernard. "Pioneering Hypar Thin Shell Concrete Roofs in the 1930s." *Beton– und Stahlbetonbau* 111, no. 3 (2016): 159–65.

Garlock, Maria Moreyra, and David P. Billington. *Félix Candela: Engineer, Builder, Structural Artist.* Princeton, NJ: Princeton University Art Museum; New Haven, CT: Yale University Press, 2008.

Hines, Eric M., and David P. Billington. "Anton Tedesko and the Introduction of Thin Shell Concrete Roofs in the United States." *Journal of Structural Engineering* 130, no. 11 (2004): 1639–50.

Mungan, Ihsan, and John Fredrick Abel. *Fifty Years of Progress for Shell and Spatial Structures: In Celebration of the 50th Anniversary Jubilee of the IASS (1959–2009).* Madrid: IASS, 2011.

SEATTLE ARCHITECTURE

Ochsner, Jeffrey Karl. *Lionel H. Pries, Architect, Artist, Educator: From Arts and Crafts to Modern Architecture.* Seattle: University of Washington Press, 2007.

Ochsner, Jeffrey Karl, ed. *Shaping Seattle Architecture: A Historical Guide to the Architects.* 2nd ed. Seattle: University of Washington Press, 2014.

ILLUSTRATION CREDITS

CHAPTER 1: BEGINNINGS

Page 15: Jack Christiansen and sister Joan, 1936. Courtesy Christiansen Personal Archive.

Page 17: Jack Christiansen, relaxing in summer 1945. Courtesy Christiansen Personal Archive.

Page 23: Christiansen's personal copy of *The Architecture of Bridges*, by Elizabeth B. Mock (Museum of Modern Art, 1949). The Museum of Modern Art, New York, NY, USA. Digital Image copyright the Museum of Modern Art / Licensed by SCALA / Art Resource, NY.

Page 24: Robert Maillart's bridge over the Arve River, Vessy, Switzerland, completed 1937. Photo taken by Christiansen in 1980. Courtesy Christiansen Personal Archive.

Page 27: Jack Christiansen during his time at the University of Illinois, ca. 1948. Courtesy Christiansen Personal Archive.

Page 30: F. W. Woolworth and Company, Chicago, 1949. Shaw, Metz and Dolio, architects. Historic Architecture and Landscape Image Collection, Ryerson and Burnham Archives, the Art Institute of Chicago. Digital File #72143.

Page 31: Hayden Planetarium, American Museum of Natural History, New York, NY. The Miriam and Ira D. Wallach Division of Art, Prints and Photographs. Photo by Morris Huberland © Photography Collection, the New York Public Library. New York Public Library Digital Collections. Image ID: 5380270.

Page 34: Jack Christiansen preparing to drive to Seattle in 1952. Courtesy Christiansen Personal Archive.

CHAPTER 2: FIRST FORMS

Page 36: Seattle waterfront and skyline, 1952. Engineering Department Photographic Negatives. Record Series 2613-07, Item 43586. Courtesy Seattle Municipal Archives.

Page 42: Details from Christiansen's personal copy of the American Society of Civil Engineers' *Design of Cylindrical Concrete Shell Roofs* (ASCE Manual 31), published in 1952. Courtesy Christiansen Personal Archive.

Page 45: Completed Green Lake Pool, Seattle. Designed by Lamont & Fey with Jack Christiansen. Christiansen's first executed shell structure, 1954. Don Sherwood Parks History Collection. Record Series 5801-01, Item 29161. Courtesy Seattle Municipal Archives.

Page 48: Seattle School District Warehouse, Seattle. Designed by John W. Maloney and Jack Christiansen. Under construction, showing Christiansen-specified movable formwork for segmental casting of barrel-vaulted shells, 1956. Courtesy MKA Slide Archives.

Page 49: Seattle School District Warehouse, Seattle. Designed by John W. Maloney and Jack Christiansen. Under construction, 1956. (*top*) Christiansen on top of cast vaults, inspecting concrete work. (*bottom*) Workers pouring concrete by bucket delivery. Both courtesy MKA Slide Archives.

Page 50: Seattle School District Warehouse, Seattle. Designed by John W. Maloney and Jack Christiansen. Complete, 1956. (*above*) View from street. Image No. 601-19. (*left*) Aerial view, 1956. Image No. 601-18. Both courtesy Seattle Public School Archive.

Page 53: Ellensburg Senior High School, Ellensburg, WA. Designed by John W. Maloney and Jack Christiansen. Arts and music building vaults under construction, 1955. Courtesy MKA Slide Archives.

Page 54: Ellensburg Senior High School, Ellensburg, WA. Designed by John W. Maloney and Jack Christiansen. Gymnasium and workshop arches under construction (arts and music building complete beyond), 1955. Courtesy MKA Slide Archives.

Page 56: Wilson Junior High School, Yakima, WA. Designed by John W. Maloney and Jack Christiansen. Gymnasium with post-tensioned concrete edge beams. (*top*) Under construction, 1956. (*bottom*) Complete, 1956. Both courtesy MKA Slide Archives.

Page 58: Jack Christiansen (*far right*) with climbing partners on top of Mount St. Helens, 1956. Courtesy Christiansen Personal Archive.

Page 59: B-52 airplane hangar for the Boeing Aircraft Company, Moses Lake, WA. Designed by Naramore, Bain, Brady & Johanson and Jack Christiansen. Single bay under construction, 1956. Courtesy MKA Slide Archives.

Page 60: B-52 airplane hangar for the Boeing Aircraft Company, Moses Lake, WA. Designed by Naramore, Bain, Brady & Johanson and Jack Christiansen. Complete, with Christiansen out front, 1956. Courtesy MKA Slide Archives.

Page 62: St. Edward Church, Seattle. Designed by John W. Maloney and Jack Christiansen. Completed 1958, contemporary view. Photo by author.

CHAPTER 3: GEOMETRIC COMPLEXITY

Page 66: Félix Candela's Cosmic Rays Pavilion, Mexico City. Completed 1951, contemporary view. Photo by author.

Page 68: Wenatchee Junior High School (now Pioneer Middle School), Wenatchee, WA. Designed by Naramore, Bain, Brady & Johanson and Jack Christiansen. Completed, 1957. (*above*) Classrooms showing undulating cylindrical concrete vaults, contemporary view. Photo by author. (*left*) Christiansen's first use of the hyperbolic paraboloid, completed. Courtesy MKA Slide Archives.

Page 69: Wenatchee Junior High School (now Pioneer Middle School), Wenatchee, WA. Designed by Naramore, Bain, Brady & Johanson and Jack Christiansen. Gymnasium under construction, 1957. Courtesy MKA Slide Archives.

Page 70: Wenatchee Junior High School (now Pioneer Middle School), Wenatchee, WA. Designed by Naramore, Bain, Brady & Johanson and Jack Christiansen. Gymnasium complete, 1957. Courtesy MKA Slide Archives.

Page 73: Mercer Island High School Multipurpose Room, Mercer Island, WA. Designed by Bassetti & Morse and Jack Christiansen. Formwork showing ruling geometry for concrete shell, 1958. Courtesy MKA Slide Archives.

Page 74: Mercer Island High School Multipurpose Room, Mercer Island, WA. Designed by Bassetti & Morse and Jack Christiansen. (*left*) Casting concrete shell and stiffening ribs, 1958. Courtesy Christiansen Personal Archive. (*below*) Complete, 1958. Courtesy MKA Slide Archives.

Page 76: Ingraham High School Auditorium, Seattle. Designed by Naramore, Bain, Brady & Johanson and Jack Christiansen. Roof framing plan, drawings dated 1958. Image No. 020-0070. Courtesy Seattle Public School Archive.

Page 77: Ingraham High School Auditorium, Seattle. Designed by Naramore, Bain, Brady & Johanson and Jack Christiansen. (*above*) Roof contour and formwork details, drawings dated 1958. Image No. 020-0073. Courtesy Seattle Public School Archive. (*right*) Aerial view of formwork and scaffolding, 1959. Courtesy MKA Slide Archives.

Page 78: Ingraham High School Auditorium, Seattle. Designed by Naramore, Bain, Brady & Johanson and Jack Christiansen. (*top*) Ground view of formwork and scaffolding, 1959. (*bottom*) Shell complete, 1959. Both courtesy MKA Slide Archives.

Page 79: Ingraham High School Auditorium, Seattle. Designed by Naramore, Bain, Brady & Johanson and Jack Christiansen. Completed auditorium, 1959. Image No. 020-04. Courtesy Seattle Public School Archive.

Page 82: University of Washington–City of Seattle Montlake Interchange, Preliminary Design, Seattle. Designed by Worthington & Skilling. Model photograph, ca. 1957. Negative No. UW37713. Special Collections, University of Washington Libraries.

Page 83, top: University of Washington–City of Seattle Montlake Interchange, Preliminary Design, Seattle. Designed by Worthington & Skilling. Sketch of typical section and elevation of pedestrian overpass, ca. 1957. Negative No. UW37715. Special Collections, University of Washington Libraries.

Page 83, bottom: University of Washington Parking Facilities, Pedestrian Overpasses and Walkways. Title sheet, 1958. Courtesy Facilities Services, Facilities Information Library, University of Washington.

Page 84, above: University of Washington Parking Facilities, Pedestrian Overpasses and Walkways. Structural details, 1958. Courtesy Facilities Services, Facilities Information Library, University of Washington.

Page 84, left: University of Washington Pedestrian Bridges, Seattle. Designed by Jack Christiansen. Under construction, 1958. Courtesy MKA Slide Archives.

Page 85: University of Washington Pedestrian Bridges, Seattle. Designed by Jack Christiansen. Completed. Courtesy MKA Slide Archives.

Page 87: King County International Airport, Hangar 9. Designed by Bassetti & Morse and Jack Christiansen. Under construction, showing movable formwork in place, 1962. Courtesy MKA Slide Archives.

Page 88: King County International Airport, Hangar 9. Designed by Bassetti & Morse and Jack Christiansen. (*above*) Internal view, 1962. (*left*) Complete, external view, 1962. Both courtesy MKA Slide Archives.

Page 91: Formwork for hyperbolic paraboloid shell, ca. 1960. Designed by Jack Christiansen and Maury Proctor for Shell Forms Inc. (*above*) On truck for delivery to construction site. (*right*) Formwork in place, ready for casting. Both courtesy Christiansen Personal Archive.

Page 92, left: Jack Christiansen standing on top of hyperbolic paraboloid umbrella, cast in forms by Shell Forms Inc., ca. 1960. Courtesy MKA Slide Archives.

Page 92, below: United Control Corporation, Redmond, WA. Designed by Paul Kirk and Jack Christiansen. Architectural diagram showing layout of hyperbolic paraboloid umbrellas, 1960. Courtesy Christiansen Personal Archive.

Page 94: Promotional brochure for Shell Forms Inc., n.d. (*left*) Cover page. (*right*) Interior page. Both courtesy Christiansen Personal Archive.

Page 95: Shell Forms Inc. warehouse structure, unknown location, n.d. Courtesy Christiansen Personal Archive.

Page 96: Church of the Good Shepherd, Bellevue, WA. Designed by Paul Kirk and Jack Christiansen. Model view, 1961. Courtesy MKA Slide Archives.

Page 97: Church of the Good Shepherd, Bellevue, WA. Designed by Paul Kirk and Jack Christiansen. Nearly complete, view from below, 1962. Courtesy MKA Slide Archives.

Page 98, left: Christiansen Residence, Bainbridge Island, WA. Designed by Jack Christiansen. Completed 1964, contemporary view. Photo by author.

Page 98, below: Mercer Island Beach Club, Mercer Island, WA. Designed by Paul Kirk and Jack Christiansen. Completed view, 1966. Courtesy MKA Slide Archives.

CHAPTER 4: AN INTERNATIONAL SHOWCASE

Page 101: Century 21 Exposition (a.k.a. Seattle World's Fair), Seattle. Aerial view of World's Fair, 1962. Record Series 9955-01, Item 73116. Courtesy Seattle Municipal Archives.

Page 103: Original Nile Temple (Century 21 Club for 1962 Seattle World's Fair). Designed by Samuel G. Morrison and Jack Christiansen. Under construction, 1956. Courtesy MKA Slide Archives.

Page 105: Lambert–St. Louis Airport. Designed by Minoru Yamasaki and Anton Tedesko. Completed 1956, contemporary view. Photo by author.

Page 108: US Science Pavilion (now the Pacific Science Center), Seattle. Designed by Minoru Yamasaki and Jack Christiansen. Precast wall panels lifted to form bearing walls of buildings surrounding courtyard, 1961. Courtesy MKA Slide Archives.

Page 109: US Science Pavilion (now the Pacific Science Center), Seattle. Designed by Minoru Yamasaki and Jack Christiansen. Openings in precast concrete bearing walls allowing passageways and light, 1961. Courtesy MKA Slide Archives.

Page 110: US Science Pavilion (now the Pacific Science Center), Seattle. Designed by Minoru Yamasaki and Jack Christiansen. Overhead arches under construction, 1961. Item spl_wl_sec_01342. Werner Lenggenhager Photograph Collection, Seattle Public Library.

Page 111: US Science Pavilion (now the Pacific Science Center), Seattle. Designed by Minoru Yamasaki and Jack Christiansen. Overhead arches nearly complete with pavilion buildings in the background, 1961. Record Series 1204-01, Item 165718. Courtesy Seattle Municipal Archives.

Page 114: US Science Pavilion (now the Pacific Science Center), Seattle. Designed by Minoru Yamasaki and Jack Christiansen. Elevated entrance slabs, under construction, showing pattern of ribs and post-tensioning cables, 1961. Courtesy MKA Slide Archives.

Page 115: US Science Pavilion (now the Pacific Science Center), Seattle. Designed by Minoru Yamasaki and Jack Christiansen. (*top*) Elevated entrance slabs nearly complete, 1961. (*bottom*) Underside of elevated entrance slabs revealing Christiansen's pattern of ribs, 1962. Both courtesy MKA Slide Archives.

Page 116: US Science Pavilion (now the Pacific Science Center), Seattle. Designed by Minoru Yamasaki and Jack Christiansen. (*left*) Complete, 1962. Courtesy MKA Slide Archives. (*above*) Aerial view, 1962. AR-07809001-ph001691. Werner Lenggenhager, State Library Photograph Collection, 1851–1990, Washington State Archives, Digital Archives, www.digitalarchives.wa.gov, February 27, 2019.

Page 118, top: International Pavilion, Seattle. Designed by Walker & McGough and Jack Christiansen. Umbrella geometry, 1961. Courtesy MKA Slide Archives.

Page 118, bottom: International Pavilion, Seattle. Designed by Walker & McGough and Jack Christiansen. Model view, 1961. Courtesy MKA Slide Archives.

Page 119: International Pavilion, Seattle. Designed by Walker & McGough and Jack Christiansen. Under construction, 1961. Courtesy MKA Slide Archives.

Page 120: International Pavilion, Seattle. Designed by Walker & McGough and Jack Christiansen. Completed view at night, 1962. Courtesy MKA Slide Archives.

Page 122: Fine Arts Pavilion (now the Phelps Center and Seattle Center Exhibition Hall), Seattle. Designed by Paul Kirk and Jack Christiansen. Architectural model, 1962. Negative No. UW 39536. Special Collections, University of Washington Libraries.

Page 123: Fine Arts Pavilion (now the Phelps Center and Seattle Center Exhibition Hall), Seattle. Designed by Paul Kirk and Jack Christiansen. (*above*) Roof under construction, 1962. Item spl_wl_sec_00903. (*right*) Complete, Space Needle in background, 1962. Item spl_wl_sec_00906. Both from the Werner Lenggenhager Photograph Collection, Seattle Public Library.

CHAPTER 5: EXPANDING AUDIENCE

Page 129: North Shore Congregation Israel, Glencoe, IL. Designed by Minoru Yamasaki and Jack Christiansen. Completed exterior view, 1963. Courtesy Christiansen Personal Archive.

Page 130: North Shore Congregation Israel, Glencoe, IL. Designed by Minoru Yamasaki and Jack Christiansen. Completed interior view, 1963. Courtesy Christiansen Personal Archive.

Page 131: Carleton College Gymnasium and Swimming Pool, Northfield, MN. Designed by Minoru Yamasaki and Jack Christiansen. Completed exterior view, 1965. Item 9008_093. Courtesy Carleton College Archives.

Page 132: Carleton College Swimming Pool, Northfield, MN. Designed by Minoru Yamasaki and Jack Christiansen. Completed interior view, 1965. Item 9008_091. Courtesy Carleton College Archives.

Page 134: Chrysler Styling Center Dome, Highland Park, MI. Designed by Minoru Yamasaki and Jack Christiansen. Completed exterior view, 1969. Reproduction No. HAER MICH,82-HIPA,2J—2. Library of Congress, Prints and Photographs Division, Historic American Engineering Record.

Page 136: Washington Corrections Center, Shelton, WA. Designed by Bassetti & Morse with Walker & McGough (Curtis & Davis consulting) and Jack Christiansen. Completed exterior view, 1961. Courtesy MKA Slide Archives.

Page 137: International Exhibition Facility, New Orleans. Designed by Curtis & Davis and Jack Christiansen. Completed view, 1967. Courtesy MKA Slide Archives.

Page 138: International Exhibition Facility, New Orleans. Designed by Curtis & Davis and Jack Christiansen. Top view of post-tensioning cables draped within the shell, 1966. Courtesy MKA Slide Archives.

Page 139: International Exhibition Facility, New Orleans. Designed by Curtis & Davis and Jack Christiansen. Segmental casting of long-span barrel vaults, 1966. Courtesy MKA Slide Archives.

Page 140: International Exhibition Facility, New Orleans. Designed by Curtis & Davis and Jack Christiansen. Worker preparing vault formwork for casting, 1966. Courtesy MKA Slide Archives.

Page 141: International Exhibition Facility, New Orleans. Designed by Curtis & Davis and Jack Christiansen. Completed side view, 1967. Courtesy MKA Slide Archives.

Page 142: Jack Christiansen (*left*) with John G. Reutter (*right*) of the Consulting Engineers Council, 1969. Christiansen received the council's Honor Award for Excellence in Engineering for the International Exhibition Facility. Courtesy MKA Slide Archives.

Page 143: Pedestrian Bridge over Interstate 5 at 195th Street, Seattle. Designed by Jack Christiansen. Completed view, 1965. Courtesy MKA Slide Archives.

Page 144: Nalley Valley Viaduct, Tacoma. Designed by Jack Christiansen. Completed view, 1971. Courtesy MKA Slide Archives.

Page 145, left: Nalley Valley Viaduct, Tacoma. Designed by Jack Christiansen. View from underneath, 1971. Courtesy MKA Slide Archives.

Page 145, right: *Building Design and Construction*, March 1970. Featuring John Skilling (*left*), Jack Christiansen (*center*), and Helge Helle (*right*). Leslie E. Robertson (not pictured) was also featured in the article. Courtesy BD+C/SGC Horizon.

Page 146: Promotional material for Skilling, Helle, Christiansen & Robertson, ca. 1975. Courtesy MKA Archives.

CHAPTER 6: THE GRANDEST SCALE

Page 148: Aerial view of downtown Seattle and waterfront, 1968. Item 77169. Courtesy Seattle Municipal Archives.

Page 151: Houston Astrodome, Houston. Designed by Hermon Lloyd & W. B. Morgan (with Wilson, Morris, Crain & Anderson consulting) and engineers Roof Structures Inc., Walter P. Moore, and Praeger, Kavanagh & Waterbury. Completed view, 1964. George Kirksey Papers, ID 1971-002. Courtesy Special Collections, University of Houston Libraries.

Page 152: Seattle–King County Stadium Report, by Praeger, Kavanagh & Waterbury, March 1966. Series 237-6-3, ARCH/DWG-17, Stadium File, 1965-1966-1967, King County Municipal Archives.

Page 154: Weyerhaeuser Wood Dome, not built. Designed by Harris & Reed and Jack Christiansen. Model view, 1967. Courtesy MKA Slide Archives.

Page 155: Weyerhaeuser Wood Dome, not built. Designed by Harris & Reed and Jack Christiansen. Diagram of construction process, 1967. Courtesy MKA Slide Archives.

Page 161: Seattle Center Stadium Design, not built. Designed by Naramore, Skilling & Praeger with Jack Christiansen. Sketch of stadium roof with Mount Rainier and Space Needle, 1969. Courtesy MKA Slide Archives.

Page 162: Seattle Center Stadium Design, not built. Designed by Naramore, Skilling & Praeger with Jack Christiansen. Christiansen's design for thin shell concrete roof with stiffening ribs, 1969. Courtesy MKA Slide Archives.

Page 164: Seattle Center Stadium Design, not built. Designed by Naramore, Skilling &

Praeger with Jack Christiansen. Sketch of potential interior condition, 1969. Courtesy MKA Slide Archives.

Page 166: Jack Christiansen underneath the Eiffel Tower, 1971. Courtesy Christiansen Personal Archive.

Page 167: CNIT Exhibition Hall, Paris. Designed by Bernard Zehrfuss, Jean de Mailly, Robert Camelot, Jean Prouvé, and Nicolas Esquillan. Photo by Jack Christiansen, 1971. Courtesy Christiansen Personal Archive.

Page 170: King County Multipurpose Stadium, Seattle. Designed by Naramore, Skilling & Praeger with Jack Christiansen. Concrete roof plan, 1972. King County Stadium Shell Segment Geometry, Series 497, Department of Stadium Administration, Kingdome architectural files, box 17, King County Archives.

Page 171: King County Multipurpose Stadium, Seattle. Designed by Naramore, Skilling & Praeger with Jack Christiansen. Concrete roof details, 1972. King County Stadium Roof Framing Plan, Series 497, Department of Stadium Administration, Kingdome architectural files, box 17, King County Archives.

Page 172: King County Multipurpose Stadium, Seattle. Designed by Naramore, Skilling & Praeger with Jack Christiansen. Architectural model, 1972. Courtesy MKA Slide Archives.

Page 178: Finite element analysis segment for computer analysis of King County Multipurpose Stadium, 1973. Courtesy MKA Slide Archives.

Page 179: King County Multipurpose Stadium, 1973. (*left*) Formwork for physical model segment. (*above*) Physical model segment for testing structural behavior. Both courtesy MKA Slide Archives.

Page 180: King County Multipurpose Stadium, Seattle. Designed by Naramore, Skilling & Praeger with Jack Christiansen. (*top*) Formwork segment, 1974. (*bottom*) Scaffolding and formwork for single segment in place, 1974. Both courtesy MKA Slide Archives.

Page 181: King County Multipurpose Stadium, Seattle. Designed by Naramore, Skilling & Praeger with Jack Christiansen. (*top left*) Four segments cast on revolving forms, 1974. Courtesy MKA Slide Archives. (*top right*) Newsletter, Issue 7, November 1974. Promotional Materials, Series 473, Department of Stadium Administration, Public Information Office, box 1, folder 21, King County Archives. (*bottom*) Exterior view, with four segments in place, 1974. Courtesy MKA Slide Archives.

Page 183: King County Multipurpose Stadium, Seattle. Designed by Naramore, Skilling & Praeger with Jack Christiansen. Aerial view, with twelve segments cast, 1975. Series 1608, Department of Stadium Administration, Tenant Services Division, Promotions/Media Relations, box 1, folder 10, King County Archives.

Page 184: King County Multipurpose Stadium, Seattle. Designed by Naramore, Skilling & Praeger with Jack Christiansen. (*above*) Aerial view, with twenty segments (one-half of the total) cast, 1975. (*left*) View from the top, nearing completion, 1975. Both courtesy MKA Slide Archives.

Page 185: King County Multipurpose Stadium, Seattle. Designed by Naramore, Skilling & Praeger with Jack Christiansen. Complete exterior view, 1976. Courtesy MKA Slide Archives.

Page 186: King County Multipurpose Stadium, Seattle. Designed by Naramore, Skilling & Praeger with Jack Christiansen. Interior, ca. 1977. Courtesy MKA Slide Archives.

Page 187: Sports fans, Kingdome parking lot, ca. 1978. Series 1608, Department of

Stadium Administration, Tenant Services Division, Promotions/Media Relations, box 9, folder 3, King County Archives.

CHAPTER 7: RETIREMENT AND PERSEVERANCE

Page 192: Shell Forms Inc. advertisement showing two options for shell buildings— hyperbolic paraboloid and cylindrical "arch," ca. 1974. Courtesy Christiansen Personal Archive.

Page 194: US Pavilion, Expo '74, Spokane. Designed by Naramore, Bain, Brady & Johanson and Jack Christiansen. Model view, 1973. Item EXPO74FF76. City of Spokane Planning Department EXPO '74 Photograph Collection, Washington State Archives, Eastern Region Branch.

Page 195: US Pavilion, Expo '74, Spokane. Designed by Naramore, Bain, Brady & Johanson and Jack Christiansen. Structural plan showing projection of overhead cable arrangement, 1973. Courtesy MKA Slide Archives.

Page 196: US Pavilion, Expo '74, Spokane. Designed by Naramore, Bain, Brady & Johanson and Jack Christiansen. Cable supported structure before covering, 1973. Courtesy MKA Slide Archives.

Page 197: US Pavilion, Spokane. Designed by Naramore, Bain, Brady & Johanson and Jack Christiansen. Completed view with fair in foreground, 1974. Item EXPO74EE97. City of Spokane Planning Department EXPO '74 Photograph Collection, Washington State Archives, Eastern Region Branch.

Page 198: Museum of Flight, Seattle. Designed by Ibsen Nelsen and Jack Christiansen. Segment of modular space frame lifted for assembly, 1986. Courtesy MKA Slide Archives.

Page 199: Museum of Flight, Seattle. Designed by Ibsen Nelsen and Jack Christiansen. Completed interior view, 1987. Courtesy MKA Slide Archives.

Page 200: Museum of Flight, Seattle. Designed by Ibsen Nelsen and Jack Christiansen. Completed exterior wall panel, 1987. Courtesy MKA Slide Archives.

Page 201: Museum of Flight, Seattle. Designed by Ibsen Nelsen and Jack Christiansen. Completed aerial view, 1987. Courtesy Christiansen Personal Archive.

Page 202: Royal Saudi Naval Stadium, Jubail, Saudi Arabia. Designed by Naramore, Bain, Brady & Johanson and Jack Christiansen. Side view of grandstand support and cantilevered overhead shell, ca. 1985. Courtesy MKA Slide Archives.

Page 203: Royal Saudi Naval Stadium, Jubail, Saudi Arabia. Designed by Naramore, Bain, Brady & Johanson and Jack Christiansen. Completed stadium viewed from a distance, ca. 1985. Courtesy MKA Slide Archives.

Page 204, above and left: Eruption of Mount St. Helens as seen from Mount Adams, May 18, 1980. Suzanne Christiansen in foreground. Photos by Jack Christiansen. Courtesy Christiansen Personal Archive.

Page 205: Eruption of Mount St. Helens as seen from Mount Adams, May 18, 1980. Suzanne and Jack Christiansen in foreground. Courtesy Christiansen Personal Archive.

Page 208: SunDome, Yakima, WA. Designed by Loschky, Marquardt & Nesholm (LMN) with Loofburrow architects and Jack Christiansen. Shell segment geometry, 1989. Courtesy Christiansen Personal Archive.

Page 209: SunDome, Yakima, WA. Designed by Loschky, Marquardt & Nesholm

(LMN) with Loofburrow architects and Jack Christiansen. (*above*) Wedge segment detail for shell roof, 1989. (*right*) Under construction, 1989. Both courtesy Christiansen Personal Archive.

Page 210: SunDome, Yakima, WA. Designed by Loschky, Marquardt & Nesholm (LMN) with Loofburrow architects and Jack Christiansen. SunDome, Yakima, WA. Complete, 1989. (*above*) Exterior. (*left*) Interior. Both courtesy Christiansen Personal Archive.

Page 211: Bainbridge Island High School Grandstand, Bainbridge Island, WA. Designed by Jack Christiansen. Typical cross-section of grandstand and overhead canopy, 1991. Courtesy Christiansen Personal Archive.

Page 212: Bainbridge Island High School Grandstand, Bainbridge Island, WA. Designed by Jack Christiansen. Overhead canopy and support detail, 1991. Courtesy Christiansen Personal Archive.

Page 213: Bainbridge Island High School Grandstand, Bainbridge Island, WA. Designed by Jack Christiansen. (*top*) Casting of canopy shells on-site, 1991. (*bottom*) Erection of precast pieces, 1991. Both courtesy Christiansen Personal Archive.

Page 214: Bainbridge Island High School Grandstand, Bainbridge Island, WA. Designed by Jack Christiansen. Complete, 1991. Courtesy Christiansen Personal Archive.

Page 216: Downtown Seattle and waterfront, 1977. NW950-75F-0-1-225. Records of the Port of Bellingham, Photographs, 1890–1999, Washington State Archives, Digital Archives, www.digitalarchives.wa.gov, February 27, 2019.

Page 220: The Seattle Kingdome and recently completed Safeco Field, 1999. Item DSCN1561, A08-065, Series 1247, Department of Transportation, Office of the Director, box 1, DVD 002, King County Archives.

Page 226: The implosion of the Seattle Kingdome, March 26, 2000. Record Series 0207-01, Item 100486. Courtesy Seattle Municipal Archives.

Page 228: The crushed concrete shell of the Seattle Kingdome, March 26, 2000. Photo by Pedro Perez / *Seattle Times*.

CONCLUSION: THE LEGACY AND FUTURE OF CONCRETE SHELLS

Page 231: Chena River Bridge, Fairbanks. Designed by Jack Christiansen. Complete, view from the water, 2003. Courtesy Christiansen Personal Archive.

Page 235: Shannon & Wilson Office Building, Seattle. Designed by Naramore, Bain, Brady & Johanson and Jack Christiansen. Completed 1959, designated Seattle City Landmark 2017. Photo by author.

Page 236: Pacific Architect and Builder Inc. Office Building, Seattle. Designed by A. O. Bumgardner and Jack Christiansen. Completed 1959, designated Seattle City Landmark 2017. Photo by author.

Page 238: Jack Christiansen in front of Bainbridge Island High School Grandstand, 2010. Photo by John Stamets. Courtesy Docomomo WEWA.

INDEX

Page numbers in italics refer to illustrations. For a complete list of Christiansen's projects, see the appendix.